T0231978

SIGNS OF MEANING IN THE UNIVERSE

JESPER HOFFMEYER

ADVANCES IN
SEMIOTICS

Thomas A. Sebeok, General Editor

SIGNS OF MEANING IN THE UNIVERSE

TRANSLATED BY

BARBARA J. HAVELAND

INDIANA
UNIVERSITY
PRESS
Bloomington & Indianapolis

En Snegl På Vejen: Betydningens naturhistorie
©*1993 by Jesper Hoffmeyer og Munksgaard/*
Rosinante, Kobenhavn
Translation © *1996 by Indiana University Press*
All rights reserved

Manufactured in the United States of America

Library of Congress Cataloging-in-Publication Data

Hoffmeyer, Jesper.
    [En snegl på vejen. English]
    Signs of meaning in the universe / Jesper Hoffmeyer ; translated
  by Barbara J. Haveland.
        p. cm.
    Includes bibliographical references and index.
    ISBN 0-253-33233-8 (cl : alk. paper)
    1. Life (Biology)—Philosophy. 2. Biology—Philosophy.
  3. Semiotics.   I. Title.
  QH501.H6213 1996
  574'.01—dc20   *96-14287*

    2 3 4 5 01 00 99 98

CONTENTS

This book deals with something for which there is, as yet, no standard word in layperson's language, though there most definitely ought to be. Please forgive me therefore for introducing, right at the outset, the word we lack: *Semiosphere*.[1]

The semiosphere is a sphere just like the atmosphere, the hydrosphere, and the biosphere. It penetrates to every corner of these other spheres, incorporating all forms of communication: sounds, smells, movements, colors, shapes, electrical fields, thermal radiation, waves of all kinds, chemical signals, touching, and so on. In short, signs of life. By the time we reach chapter 5 we will be in a position to go into this word in depth.

PREFACE

Every organism on Earth is tossed at birth into this semiosphere, to which it must adapt correctly if it is to survive. In recent decades we have become increasingly aware that everything in this world is connected; that, for example, fungicides used in southern Europe will, by the strangest routes, come to contribute to the ultraviolet radiation to which innocent Lapps in the north of Norway are exposed. We now know that all of the creatures on this planet exist as part of one, great ecological community, the biosphere.

Oddly enough, though, only very recently has it begun to dawn on us that this term, the *biosphere*, does not—as we had thought—imply holism, the universal view of things. When the concept of the biosphere was introduced the semiosphere had not yet been thought of. Our ecological awareness is still stuck at the physico-chemical level where energy currents, biomass, and food chains constitute standard categories. Consequently, we tend to overlook the fact that all plants and animals—all organisms, come to that—live, first and foremost, in a world of *signification*. Everything an organism senses signifies something to it: food, flight, reproduction—or, for that matter, despair. Because of course human beings also inhabit the biosphere.

Very briefly, in the ten chapters that follow I intend to show how we humans live, like all other animals, plants, protists, fungi and bacteria, within a semiosphere. And that the biosphere must be viewed in the light of the semiosphere rather than the other way around. I will follow the growth of this semiosphere from its infant beginnings around seven hundred thousand years after the big bang right up to the animals and plants of today. Not only that but I will follow the semiosphere into the heart of organisms, to where cells swarm around amid a cacophony of messages. And I will demonstrate how it was possible for these swarming cells finally to turn into thought swarms within human beings who knew how to talk to one another and could differentiate between good and evil.

The key question lying at the root of all this is: How could natural history become cultural history? Or, to put it another way, how did Nature come to mean something to someone? In fact "someone" could be said to be the central character of this book. How did something become "someone"?

It has taken me a good number of years to develop this "semiosphere theory." And in the process I have drawn inspiration from many quarters. My long-standing collaboration with Claus Emmeche has proved to be of particular, and vital, importance. But a lot of other people have helped and encouraged me, whether by challenging my arguments, pointing out flaws, making suggestions or affording fresh perspectives. I wish to thank all of them. And my especial thanks to Myrdene Andersen, Claus Bahne Bahnson, Niels Bonde, Bodil Bredsdorff, Peder Voetmann Christiansen, Mogens Clæssen, Kjeld Fredens, Nils Gunder Hansen, Jesper Hermann, Jörg M. Hermann, Kaj Frank Jensen, Mogens Kilstrup, Simo Køppe, Ejvind Larsen, Svend Erik Larsen, Benny Lautrup, Tor Nørretranders, Stanley Salthe, Thomas A. Sebeok, Pia Skogemann, Frederik Stjernfelt, Ole Terkelsen, Thure von Uexküll, and Jon Wetlesen. And finally, my thanks to Irmelin Krasilnikoff for her indefatigable help in retrieving information and to Merete Ries for her constant support and enthusiasm.

It goes without saying that none of these people shares the slightest responsibility for the final text.

In this connection I should just mention that ever since reading Boris Vian's book *Froth of the Day* back in the sixties I have cherished a provocative dream of making one sentence from his foreword my own: "Everything on the following pages is absolutely true, as I have made it all up from beginning to end," he writes—as I recall it—so forthrightly.

But it has to be said that the charm of this brazen subjectivity faded considerably once I discerned that it was merely an empty protest against the scientific world's every bit as brazen idolization of objectivity. So I had better admit that my own contribution to this book is very modest. By far the greatest contribution has been made by reality—or, to be more precise, scientific reality.

The problem is that, to some extent, all of the sciences have their own "reality." And in a way what this book is saying is that it is just not good enough for each scientific discipline to content itself with its own "reality"—even when this reality is actually irreconcilable with the reality of other sciences.

To be decent scientists we must take one another's "realities" seriously enough to try to eliminate the contradictions. *Signs of Meaning in the Universe* suggests one way of doing this. There may, of course, be other ways.

On lumps in nothingness, on "not"

"In a random spot on an insignificant planet in a far-flung solar system there emerges a creature born to be of especial worth. What an absurd idea!" Thus, in his major thesis[2] on the relationship between theology and science, Danish theologian Viggo Mortensen sums up one of the problems of existence: as human beings we cannot help but feel that we are of some significance, that we are worth something. Unfortunately, however, science does not provide us with the remotest justification for feeling thus.

One can of course choose to side with science and gainsay humanity's claim to any particular signification or worth. One could, for example, assert that this feeling that human life does signify anything is an illusion on a par with the illusion that the sun rises. In the sixteenth century Copernicus appalled our forefathers by proving that, in reality, the sunrise resulted from the fact that we ourselves—that is to say, the Earth—were spinning round like a top. Is this feeling of humanity's worth not a similar illusion? A projection of our own stupidity onto a universe that has never done us any harm?

Fortunately ideas tend to take a pretty relaxed attitude to reality. If we were to demand that ideas drag out their lives as hostages to reality, as we ourselves must do, then we should never be able to make use of them, either for fun or for creating anything new. Because before being created new inventions must, necessarily, be unreal. That is why we say there is no tax on thinking. The imaginary overstepping of all boundaries is quite permissible. No toll fee need be paid until they are overstepped in reality.

Similarly, however, we must be allowed to feel skeptical of what we call daydreaming. Anyone who denies the significance of humanity will soon come to grief in reality. He may keep his ideas to himself, but as soon as he starts to express them he is skating on thin ice. After all, what do his words purport to mean? And on a more practical note: deny the significance of human beings and one will find it hard to justify the very aspects of civilization that have cleared the way for the enlightened principles invoked in one's denial: freedom of thought and human rights. Since human rights are inalienable and apply quite specifically to everyone, they can be justified only by the single, unadorned fact that we are human. And if that is to be seen as justification, then humanity must possess particular worth, in some sense or other.

Here, rather than contest the validity of humanity's "worth" or significance, we are going to undertake an investigation into the other side of the absurdity, that "insignificant planet." Just what right does science have to claim that this planet is a place of no significance?

Not that long ago a Danish newspaper reported that the American COBE satellite—in lone orbit around the Earth since 1989—had picked up "the echo of the big bang" which, current theory has it, is supposed to have produced the Universe out of absolute nothingness at some point just under fifteen billion years ago. For some time now we have known of the existence of traces of the vast, energy-rich sea of light that came into being approximately seven hundred thousand years after the big bang. Today, these traces are present in the form of cosmic background radiation, the wavelength and temperature of which reveal exactly how it originated. What are new however are the irregularities that have been discovered in the initial surge of radiation. The sensitive satellite instruments succeeded in registering a temperature variation in the radiation of just thirty millionths of a degree! This variation, infinitesimal though it may be, has been interpreted by scientists as "ripples" or "lumps" in the original waves of radiation—a sort of scar, one might say, left by the big bang. According to Dr. George Smoot, an astrophysicist at the University of California at Berkeley, this discovery provides us with proof of the birth of the Universe.

Now the reader may well think that it is hardly necessary to produce proof of the birth of the universe. We must all be prepared to believe that the universe exists since, if we did not, we could hardly debate the point. But, logically speaking, there is of course the other alternative: that the universe was never born because it was there all the time. If, in this case, the bang and the birth are one and the same thing, then George Smoot has in fact come up with evidence, of a sort, that the universe has only been around for a limited number of years. And compared to eternity fifteen billion years is a limited space of time.

But this "scar tissue" which has been discovered in the cosmic background radiation is also important because it explains why there are lumps, i.e., stars, planets and galaxies, in the universe. And should the reader now ask why on earth there should not be lumps in the Universe, then the answer must be that even lumps require an explanation. For as we have learned, in the beginning there was nothing. But then this nothing exploded and became something, namely a cosmic sea of light bursting with energy, one which has taken fifteen billion years to expand to its present size. So far so good, but why was the matter not evenly distributed? Why is the same amount of matter and energy not found everywhere throughout the Universe?

I am making a particular point of spelling out these questions because the answers to them are, I believe, of profound importance to the question addressed in this book: how can signification arise out of something that signifies nothing? Because the question is, in fact, whether or not signification has its beginnings as something in the nature of "lumps in nothingness." But more of this later.

I found it therefore quite disconcerting to read in this newspaper article about the "echo of the big bang." I have, as it happens, no trouble in accepting the idea that once we have discovered scars and thereby established the existence of wounds—i.e., ripples in the anticipated evenness—in this cosmic sea of light, then it also becomes possible for us to say that these ripples consolidate into stars and planets. This is the kind of development process in which the so-called butterfly effect has taught us to believe. The butterfly effect is a humorous illustration of the impossibility of predicting the development of complex systems. Where the weather is concerned, we have to be prepared for a butterfly fluttering around in Beijing today to have an effect on storm fronts in

New York in a month's time.[3] In his book on "chaos theory" James Gleick illustrates this effect by means of an old rhyme:

> For want of a nail, the shoe was lost;
> For want of a shoe, the horse was lost;
> For want of a horse, the rider was lost;
> For want of a rider, the battle was lost;
> For want of a battle, the kingdom was lost![4]

But, again, where did the scars come from in the first place? Evidently the astrophysicists cannot help us there: "One question to which COBE cannot provide a conclusive answer," writes Danish science correspondent Jørgen Steen Nielsen, "is *what* caused these ripples to occur during those first split-seconds of the Universe's life. What unlikely event created the conditions that led to our being here now, able to sit in our corner of the Universe and look back on the beginning of all things."[5]

To my astonishment Nielsen, whom I have always found to be a first-class journalist, then goes on to make the following ridiculous statement: "And the point on which COBE leaves science not one whit the wiser is this: what was there *before* the big bang? *What was this "nothingness"* that exploded and created space, time, and matter, almost fifteen billion years ago?" (My italics.)

If the word "nothingness" means anything, it must mean the absence of anything whatsoever, including the absence of any significance whatsoever. In my vocabulary there can be no difference between the one "nothingness" and the other "nothingness." Well, where would the difference lie? But even the absurdity of asking "what sort of nothing" we are dealing with here might well steer us toward an interesting discovery. Is it not after all the case that the thought of "nothingness" is quite mind-boggling: an almost impossible concept, and one which science is now being urged to eradicate? Or is science itself doing the urging? In short, we *want* there to be lumps in this nothingness. We instinctively load it with significance. The abhorrence of a vacuum nestles deep within all human beings.

Envisaging absolute nothingness is a *logical maneuver* that human beings are born to master. This maneuver lies hidden within the tiny word "not." Logic simply demands that if we can picture something that exists, a snail on a path for example, then we must also be capable of imagining the possibility that there is not a snail on the path. Hence, if we can envisage the universe, we must be able to envisage the concept of absolute nothingness.

If we then ask what this "nothingness" is or was we are actually denying our denial and to some extent re-creating the universe. Not-something becomes not not-something which, it follows, must be something. Interestingly enough, one particular cosmological theory does operate with just such a sequence of negations, whereby the interminability of the universe is resurrected in the form of an endless repetition of the sequence big bang, expansion and contraction. And from there the question arises, quite logically, as to whether there was not an initial big-big bang, and so on and so on.

But to me a cosmology that wishes to place the dawn of the universe within a nothingness that cannot be comprehended as anything other than a mental exercise, a logical maneuver involving an abstract notion of "everything-ness," seems a rather dubious proposition. Taking it seriously must inevitably bring with it the fear of inadvertently hatching a new universe every time we conceive the idea of nothingness. Because only by being conceived of can nothingness exist. Mind-boggling indeed.

Nothingness forms a fitting counterpoint to the central theme of this book—signification. But before going on to discuss the conflict that exists between these two, we will have to take a closer look at what this term "signification" actually means.

As far back as 1952, the visionary thinker, anthropologist and communications scientist Gregory Bateson got to the very heart of our problem. He was fascinated at that time by an odd question: Do jackdaws know what they are cawing? In one of Konrad Lorenz's works, Bateson had read that jackdaws cry "kiaw" when about to turn for home. If this cry is merely what the jackdaws' frame of mind bids them utter, then the story ends there. But what if the jackdaw "knows" that it is transmitting a signal, a signal that means something along the lines of "let's head for home, boys"? In that case, surely there must be a good chance that some jackdaws might cheat—by transmitting the signal just for fun, for instance? Or that the signal might not always be obeyed as befits a law of nature?

Either the jackdaw caws because it can do nought else (as Martin Luther is supposed to have said), and then we are talking of laws of nature (or the will of God). Or it caws because it feels that it is *in charge* and has decided to start the journey home. In which case, its cry has some *signification*. In any event, thought Bateson, it ought to be possible to determine by means of research whether any jackdaws disobeyed or cheated.

But Bateson left the jackdaws to get on with being jackdaws, and instead took himself off to a San Francisco zoo, where he proceeded to decipher the secrets of the monkey house. What he observed there constituted a major breakthrough—not because he was the first to have observed it, but because he was so accurate in his estimate of its implications. The monkeys were engaged in so-called play, i.e., an activity in which they exchanged signals similar to those seen during combat. But to a human observer it was quite obvious that there was no talk of combat here *and also* that the monkeys themselves did not regard it as such.

He saw how the monkeys snapped at one another while creating an imaginary combat situation. According to Bateson, this snap actually constitutes the following "meta-message": "This is not a bite."[6] The absence of a bite is announced by the presence of the snap. The snap is an indication of something which is *not* there.

While it might be that only very few people would consider "lumps in nothingness" as having anything to do with signification, most of us would surely grant that the snap with which we are dealing here does in fact have some kind of signification. It refers to something other than itself, whether the monkeys are aware of it or not. Evidently even we humans are going to have to come to terms with the fact that for us by far the majority of everyday phenomena remain unconscious acts. This, despite the fact that they can be of great significance and could even cost us our lives, should this significance not be heeded—as, for instance, with the signals transmitted by traffic lights. It is a well-known fact that we often obey these signals quite automatically, while our thoughts are somewhere else entirely, both in time and in space.

Gregory Bateson's radical theory states that this "play," with its imaginary bite, constitutes a modest first step in the evolutionary process which led to the development of human language. This theory breaks with the belief that the origins of spoken language must be sought in body language. Body language is an attribute common to both animals and humans, and there is no evidence that human beings "preferred" to use speech. More likely, the spoken word is a completely new invention which has been spread on top of body language like a layer of whipped cream. Human beings are every bit as capable as animals of communicating with their bodies; just think of dance or mime—not to mention telegenic politicians. Body language is obviously not a communicatory appendix on the brink of extinction. But such extravagance, equipping an animal with two means of performing the same function, is from a biological point of view unheard of. The only sensible conclusion, Bateson maintains, must be that the spoken word communicates something quite different from the body language. In a nutshell, Bateson's idea is that speech creates a distance which allows for an absence or, as it were, for a "not"; that it is this distance which in its most primitive form was established with the monkeys in their "play," the ritualized indication of an absence.

Body language can refuse (by a shake of the head), or it can dismiss (e.g., with a shrug of the shoulder), but it cannot deny. It cannot express the idea that it is not snowing, that it is not raining, that there is no snail, or that there is no country called Spain. (While the sign language employed by the deaf can express denial, this is not true body language but a re-coding of the spoken word.)

Strangely enough, this very facility to think of, or talk about, something that does not exist holds the key to all manner of inventions. It is precisely through this logical exercise, whereby we imagine something other than what is actually there, that imagination is given free rein. It becomes possible to conceive of an infinite wealth of possible worlds. Let the tax-free thinking game commence.

That imagination depends upon an alienation, a denial—in other words in the casting loose of humanity's indubitable moorings in the existing world—may seem to belie the idea of intuition. And this view is, in any case, at odds with the widespread belief that losing ourselves in the primitive rhythms of African or Latin American dance is a singularly effective way of unleashing our innate creative force. Or the notion that an unspoilt child, still living "in tune with its inner being," should have some superior direct link with the wellspring of creativity.

Faced with this myth of idyllic childhood it might perhaps be appropriate to cite the French psychoanalyst Jacques Lacan's view of the earliest stages of childhood as being a state of "insanity" in which the child confuses images with reality, fantasy with its perception of the world around it—in short a state characterized by hallucinations. When fantasy and reality are one and the same, fantasy is not fantasy but reality.

In Lacan's analysis, between the ages of six months and eighteen months the child goes through a so-called mirror phase in order to form an ego. It is through its contact with another person, that is to say by seeing itself reflected by its mother or, generally speaking, "the other," that the child is able to form an image of itself. "The other offers the child a form or a gestalt that provides an alternative to the chaos which the child is, at this point, experiencing. The child thus forms an image of itself via the other, through what, in psychoanalysis, is termed primary identification. Consequently, this process also involves an alienation or a denigration of the child's self, since its self is just what it does not encounter, but an image of itself reflected in the other."[7]

According to Lacan, then, the price one pays for experiencing oneself as a complete entity, distinct from all others, is this split between the subject and the image of the subject.

The idea that a fundamental existential split is inherent in human nature is a recurring theme in the work of those philosophers, psychologists, and psychoanalysts who have devoted themselves to the study of "the self."[8] Thus the shaping of the subject in childhood seems to repeat the very process which, Bateson suggests, holds the key to the evolution of the human race: an alienation process based on the creation of an imaginary reality, a reality which is not real.

And of course this split also holds the key to human desire. The longing to be made whole again that is life itself: the endless investigation of all the manifold themes of existence. It is this split, this fundamental yearning, that endows the world with *signification*, that makes us desire it. We do not desire what we have. We only ever desire that which we could possibly—or possibly not—have. The signification depends on something which is not itself, on a schism within "something" (e.g., externally) which has some relation to "something else" (e.g., internally). It is from our alienation that our longing derives; our homing instinct. The holistic dream of eliminating this split is a dream of blissful death, of the end of all signification.

In his research into communication Anthony Wilden has shown that in essence the word "not" merely constitutes a rule as to how one performs an either-or operation.[9] But, since either-or operations are implicit in everything human beings do or think, it is a pretty crucial rule. Because only by differentiating can we ever hope to cope with life. Consciously or unconsciously we are constantly having to slot the phenomena around us into categories: this is coffee, this is a falling tree, this is a high-pitched wail denoting the imminent appearance of a police car. Each time, the same thing happens: one element (the coffee, the tree, the wail) is isolated from everything else (the background); we create a gestalt. In other words, we employ the rule known as "not": this is not background, this is a tree that is about to fall on my h . . .

So the "not" rule is the very first requirement for making any sense of this world. And if we then look more closely at what lies behind this "not" rule, we will see that we are dealing with something quite fundamental. Wilden illustrates this with the following figure.[10]

This figure depicts a gestalt: a white hole (A) on a gray background (B). With this figure we have, in fact, made a differentiation between something which is B and something which is *not* B but A. Any given

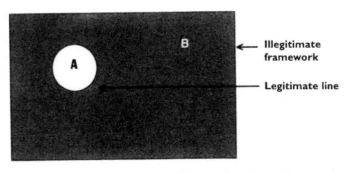

FIGURE 1. *Gestalt Diagram showing illegitimate framework and legitimate line.*

point in the figure is either A or B. Later, we will refer to this as "digitization": making a hole in something that is continuous. At this stage, however, that is not so important.

Wilden's point is that the word "not" corresponds to the circumference of the hole, the boundary between hole and not-hole. "Not" is a boundary. This boundary, the circumference, is unique because it exists nowhere but in the mind of the one who has pictured it, the observer. The boundary, or—to use Bateson's term—the difference,[11] is neither part of the hole nor part of the background; it is, in fact, a mental exercise. It forms the very roots of signification. Or, to put it another way: the boundary is not a part of the world unless "someone" chooses to picture it.[12]

And in a sense what or who this "someone" might be is exactly the question posed in this book. Who is capable of making "lumps in nothingness"? When did it start? And to what did it lead?

But we cannot leave Wilden's figure without mentioning that it contains two lines. One between A and B and the other between A and B on the one side and the rest of the universe on the other. The problem is that we cannot draw the first line without, somewhere, drawing the other. "Even if we think," writes Wilden, "that we have successfully divided the whole of reality and unreality into two sets by drawing a line between A and non-A ... the act of drawing that line defines at least one system or set as belonging neither to A nor non-A."[13]

This system is "someone." The framework therefore represents a goal-directed system beyond the system. Wilden calls this framework logically illegitimate because it belongs to a different logical type than that of the figure.

In our attempt to establish who "someone" might be we shall, in the next chapter, try first of all to show that "someone" must at least be capable of forgetting.

On history and codes: The dialectic of oblivion

On the face of it, in modern society beauty might seem to have become something of an affliction. The cult of the body, of looking good, is pursued with a devotion that would, a few decades ago, have been labelled neurotic. As one plastic surgeon recently told the New York Times: "In a tight job market people look upon a face-lift as an investment in their own ability to compete."

Strangely enough it is the young who suffer the worst of beauty's torments, even though one would expect them to be the most blessed in that area. For young people do not really stand a chance. If they do not conform to the contemporary ideal of beauty, cosmetic remedies are their only hope.

As we grow older, matters improve. We acquire wrinkles and mannerisms; habits, good and bad, take root. The stories of our lives assume bodily form; our autobiographies are there in our faces for all to read. And if there has been some beauty in any life story, then perhaps that will shine through. Hence old people can possess a beauty that no young person could ever hope to match. But then maybe it takes an old person to see it.

In a sense growing old also means becoming worn. And people with physically tough or monotonous jobs will naturally become more worn than others. And yet "worn" seems far too inadequate a word to describe the physical changes wrought by age. It would be far more true to say that traces of the past are retained in a way that reflects the course of a life: one might say the body remembers.

Take, for example, the way we become immune to an infectious disease because our bodies contain certain cells that remember the bacteria of the disease in question and are able to initiate the production of antibodies; or the way we contract an allergy because our immune system has recognized some alien substance to which we have frequently been exposed and reacted by taking excessive defensive action. But our bodies also harbor memories, of a sort, of our psychic lives. Psychological tension becomes more deeply entrenched in the form of physical tension, lost or suppressed memories can suddenly be triggered off by physical therapy, and psychological traumas may lead to paralysis.

I consider the term "to remember" appropriate in this context because what we are talking about here is a heavily selective retention of past events. Fortunately, the one thing the body is best at is erasing its memories, forgetting. Not everything is remembered, only those things that are of significance. Minor wounds heal, tiredness is dispelled by sleep, my own nervous tension is eased when a nice fat check drops through the mailbox, stress levels fall and stomach ulcers cease their gnawing once an exam is over, the apathy of a broken heart is conquered by love of life. Time heals—almost—all wounds.

That, however, is not how it works with the inanimate side of the natural world. The poor Moon, for example, has literally no earthly chance of healing the wounds it suffered during the meteor storms of the past. These same meteor storms did of course hit the Earth too but, while the scars on the Moon can be viewed with the naked eye, it would take a real expert to find many traces of meteor craters on Earth today. The Moon cannot forget and, since it cannot, it would seem rather far-fetched to describe its meteor craters as memories.

On Earth the flow of water causes constant erosion and this, over millions of years, can erase all tracks. But surely even that could not be termed forgetting? In any case this process cannot be linked to anything specific. What is erased by erosion and when this takes place are determined by the thoroughly universal laws of physics. And though traces of the past may be wiped out more quickly here on Earth, still they are often retained for a good while. Fossils embedded in cliff faces and the contours of a moraine landscape remind us to this day of events that took place in the dim and distant past.

Here we would appear to be faced with a most crucial difference between the living and the lifeless. All living systems are fragile and in principle everything is forgotten once it dies. Nevertheless, thanks to that ingenious process known as procreation, an inheritance is bequeathed to posterity. This inheritance is a very sophisticated phenomenon, the essence of which is seldom properly explained. The essence of procreation lies in a principle which we will call coding—or, even better, *semiosis.*

This term is utterly central to the thinking behind this book and we shall shortly be returning to it. For the moment, however, let us sum up the principle of this inheritance by saying that it represents a kind of charting of the race's experience for the purpose of surviving under the given conditions. One might say that this inheritance *testifies* to the past, though it does not carry it around the way the Moon bears the scars left by meteors of the past. And, as with all witnesses, what this inheritance passes on can be no more than an unreliable and inadequate account of what has happened. Nonetheless, thanks to this process every single life-form in existence today has, lodged inside its genetic material, the sinuous trail of its evolutionary past harking all the way back to the dawn of life—while it is itself busy incorporating the experiences of today into the future.

This evolutionary trail is clearly detectable in the development of the human embryo. Thus, at an early stage, the fetus develops a structure corresponding to that of the gills of a fish, later on a three-chambered heart of the type found in reptiles is formed, and later still a mammalian tail appears. A living creature as history made flesh.

A couple of million years ago South America and North America collided with one another as the result of the continental drift. At that time they had been totally separate from one another for seventy million years. South America had, for a spell, abutted on Australia and was inhabited by a mixture of marsupials and mammals. The marsupial population included a saber-toothed tiger which with its huge, overgrown canine teeth and stunted molars seems likely to have lived as a "vampire," preying on large thick-skinned animals such as hippopotamuses, elephants, and the like. "Its long dagger-shaped teeth made it possible for the saber-toothed tiger to sever its prey's carotid artery, after which it could lap up the blood. Its teeth seem to have been singularly unsuited to tearing and chewing flesh."[14]

Now, with the two continents connected by a narrow isthmus (later to become Central America), this South American marsupial saber-

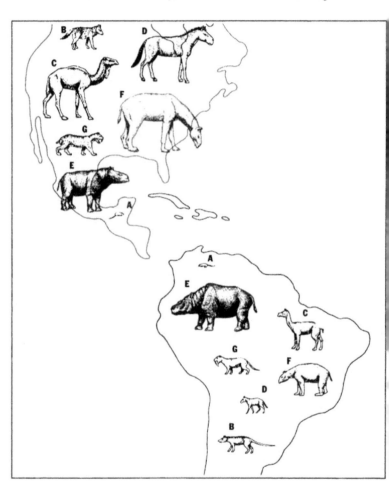

FIGURE 2. *Sample of the fauna of North and South America prior to the establishment, around two million years ago, of the Central American land bridge. G denotes the saber-toothed tiger. Three South American marsupials—A, B and G—are shown. Taken from May.*[15]

tooth suddenly found itself confronted by a North American saber-tooth which, outwardly at least, did not seem all that different from itself. The North American saber-tooth was, however, not a marsupial but a mammal. And the closest common relative of these two big cats had lived one hundred and twenty million years before. This common ancestor, a little creature a bit like an opossum, dated from the start of the Cretaceous era. All mammals—marsupial or placental, swimmers or burrowers, runners or climbers—are descended from this creature. And the saber-toothed tiger is one of the very last sprigs on the ancestral tree, designed to prey on the biggest and most thick-skinned of all mammals.[16]

Each of these two animals, by its own 120-million-year detour, had wound up with the very same specialization: that of constituting the carnivorous link in the food chain which stops at the largest of the herbivores. Despite looking alike these two animals were in fact separated by a total of two hundred and forty million years of history. And yet nothing had been able to erase the original difference between them: while at a very early stage of the embryo's development the one saber-tooth had managed to establish those nerves and muscles necessary for it to crawl out and into a little protective pocket, there to cling to its mother's nipple, the other saber-tooth stayed safe inside "Mummy's tummy" and waited to be born.

The meeting of North and South America condemned these two family histories to compete through their existing exponents. In this case both histories lost. First the South American became extinct, to be replaced by the North American species. And then somewhere around 10,000 years ago that species too died out. The saber-toothed tiger no longer walks the Earth but it seems very likely that humanity had a large part to play in this particular memory lapse.

The knack of forgetting holds the key to life's knack of incorporating the present into the future. It is precisely because living systems carry out a selection process, forgetting somewhat more of what is "unimportant" than of what is "important," that we can talk about memory. This is possible because all life is based on one fundamental schism.

Every life form exists both as itself, i.e., as an organism of "flesh and blood," and as a coded description of itself, the latter being lodged within the remarkable DNA molecules of which the genetic material is composed. In short, the genetic material contains a coded version of the organism, almost in the same way as a recipe from a cookery book contains an evening meal in code. This comparison is not as far-fetched as it may sound. Faced with the complex challenge of embryogeny, of having to divide and sub-divide trillions of times in order, at last, to become a kitten or a foal or a baby, the fertilized egg has to obtain the necessary instructions from somewhere. In the cook's case the instructions are provided by the recipe. And in exactly the same way the egg has a recipe tucked away inside its genetic material. This is simply a coded description of how to construct an organism, and of all the creatures on this Earth only the fertilized egg is capable of reading this code and translating its message into an organism.

So we can say that what is alive, the organism, is different from what survives, the genetic material. It is the coded version, the genetic material, that is passed on to the next generation by means of procreation, while the organism must die. So what survives is in fact a code for something else, an image of the subject—not the subject itself. Life is survival in coded form.

The tremendous dynamic force triggered by this schism is something we will examine in more depth in chapter 4. Here we will confine ourselves to noting that every single generational change that occurs by way of this translation ploy involves a rewriting of the species. Procreation ensures that a fortunate number of those coded versions which formed the accumulated genetic material of previous generations are brought together and mixed into new patterns. When we reproduce ourselves our self-descriptions become intermingled. In other words we bequeath a biological account of ourselves to posterity.

The time has now come to introduce yet another of this narrative's heroes, the American scientist and philosopher Charles Sanders Peirce, who lived from 1839 to 1914. Peirce was one of those catholic thinkers who are rarely held in any high regard by their contemporaries. The full significance of his life's work is only now being recognized and volume after volume of his work is now gradually finding its way into print. Of the 80,000 odd pages left behind by Peirce many were written in an out-of-the-way attic room which—legend has it—he rendered inaccessible to his creditors by pulling up the ladder.

Peirce's theories on the fundamental ordering of the world (his metaphysics), to which we will return in the next chapter, certainly swim against the tide, even of modern scientific thought. Not that this should discourage anyone from taking a look at Peirce's work since, no matter what might be said in Science's favor, it wins no special prizes for profound soul-searching. To quote the scientific philosopher Imre Lakatos: "Scientists know no more about science than fish know about hydrodynamics." But on this point Peirce does have something to offer.

In this context, where we are tackling the question of heredity and forgetfulness, Peirce's logic stands us in good stead. "I wish to reason in such a way that the facts shall not . . . disappoint the promises of my reasoning . . . I ought therefore to present my reasoning in such a way as to avoid such surprises."[17] And here, of course, we have the justification for taking an interest in logic, the study of whatever denotes valid thought. Peirce goes on: "Every reasoner has some general idea of what good reasoning is. This constitutes a theory of logic."[18]

The great thing Peirce perceived was that any form of logic which is based on two-factor, *dyadic,* relations is too limited. Bound as it is to the single dimension of the linear chain, it cannot be made to branch out (fig. 3).

He believed that logical processes ought rather to be regarded as a multi-dimensional network. Such a network can be arrived at by combining three-factor relations, *triads* (fig. 4). Thus, in Peirce's work, valid thought always presupposes a relation between three things. This could, for example, be cause and effect *plus* the observer who connects these two. Nothing, however, is achieved by working with relations between four or five factors since these can always be broken up into three-factor relations (fig. 4B).

The consequences of such a shift from dyadic to triadic logic are far-

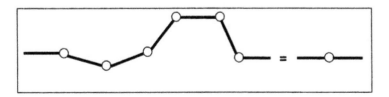

FIGURE 3. *Two-factor (dyadic) relations cannot be made to branch out.*[19]

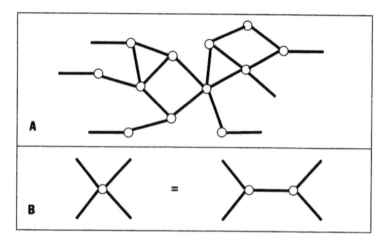

FIGURE 4. A. *Logical processes constitute multi-dimensional networks. B. A four-factor relation can be produced by connecting two three-factor relations (triads).*[20]

reaching. First and foremost, they lead straight to the realization that one cannot, in all conscience, ignore the fact that there is always "someone" (e.g., the observer) to draw the inferences which we hold to be true. Peirce was deeply aware that logic is, at heart, the philosophy of communication or, as he himself said: "All thought being performed by means of signs, logic may be regarded as the science of the general laws of signs."[21] If anyone has any doubts about thought being performed by means of signs, I suggest that you think of an elephant. Then ask yourself whether there was an elephant inside your head while you were thinking. It is to be hoped that the answer is "No."

Peirce called the triad—the basic relational element in logic—quite simply a sign. The Greek word for sign is *semeion* and so Peirce called his logical theory *semiotics,* the study—in the broadest sense—of signs. And the actual sign function—the process by which signs are exchanged— is known as *semiosis.* In this book the word "semiotics" refers to the Peircean school of sign theory. The other main school of thought within semiotics, the Saussure school, does not lend itself in the same way to integration into a biosemiotic study. It took this author a whole year of agonizing to come to this conclusion.

Peirce himself defined a sign in the following, somewhat cryptic, fashion, which I will attempt to clarify below: "A sign ... is something which stands for something to somebody in some respect or capacity."[22]

Take, for example, the case of a small child who suddenly breaks out in a rash of red spots. The mother takes the child to the doctor who establishes that the child has measles. To him the red spots are a sign of measles. But to the mother they are merely a sign that the child is sick. So the red spots are not automatically a sign of measles to just anyone, but only to "someone," namely the doctor. We can depict this connection as a triad (fig. 5A) which thus becomes a specific example of Peirce's general sign triad (fig. 5B).

In the general instance then, the sign represents a relation between three factors: (1) the primary sign—the sign vehicle—i.e., the bearer or manifestation of the sign regardless of its significance (e.g., the red spots); (2) the object (physical or nonphysical) to which the sign vehicle refers (e.g., the illness, measles); and (3) "the interpretant" i.e., the system which construes the sign vehicle's relationship to its object (e.g., the mental processes in the physician's head). To be a sign in Peirce's sense of the word all three of these elements must be present.

As we can see, such a relationship does exist in the case we are investigating here: Evolutionary forgetfulness. The selective incorporation of the present into the future can be defined as a combination of two semiotic processes reaching out to one another: embryogenesis and procreation.

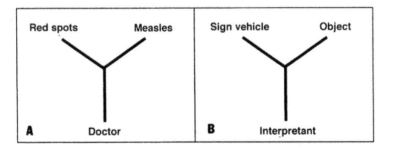

FIGURE 5. A. Red spots—Measles—Doctor. B. Sign vehicle—Object—Interpretant. A. Red spots as a sign of measles. B. The general sign triad.

Embryogenesis, or ontogenesis, is a sign operation in the sense that a one-dimensional "DNA inscription" containing—as mentioned above—"a coded version" of its parents is converted into a three-dimensional organism "of flesh and blood." The genome (the sum total of an individual's genetic material) is therefore a sign vehicle, or even better: a set of sign vehicles, referring to the construction of an organism, the ontogenetic trajectory. The question is, for whom? Who, in this case, is the "someone" who can interpret the signs?

The answer is that if the genome does not happen to be inside an egg cell then nothing at all will happen. The genome cannot put together a chicken all by itself, any more than a cookbook could roast one. An inscription is not meant to be active, it is meant to be deciphered, and so it is with the DNA inscription. We must therefore conclude that it is the fertilized egg cell which is responsible for the deciphering or interpretation. As the egg gradually interprets the genome it splits up into billions of cell lines, becoming, in other words, an organism. The egg—or if one prefers, the growing embryo—is the "someone" for whom the genome represents a sign of a specific process of development. Which gives us the triadic semiotic relation presented in figure 6.

It might be helpful, at this point, to give an example: for the greater part of its life the dreaded locust is just an ordinary, harmless grasshopper. Only when climatic conditions force grasshoppers to congregate within a limited area is a totally new set of dynamics brought into play. This mechanism works in the following way: along with their excreta

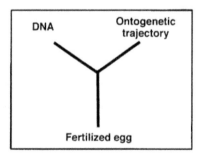

FIGURE 6. *DNA—Ontogenetic trajectory—Fertilized egg. The fertilized egg understands the DNA message. That is to say, it interprets it as an instruction to construct the organism and thus implement the ontogenetic trajectory.*

the grasshoppers release a particular chemical compound, a phero-mone, and as the density of the grasshopper flock increases, so too does the concentration of this compound. This increased concentration functions as a sign to all the fertilized eggs. The eggs then proceed to form an entirely different type of grasshopper—a migratory grasshop-per, a locust. Hence, the same DNA is interpreted in a different way. The locust's wings and flying muscles are more powerful and its life cycle accelerated. Added to which, it produces a substance that attracts other locusts. These locusts then split up into enormous flocks to fly many thousands of kilometers, devouring every green thing in their path.

So here we have external forces prompting a switch to an alternative form of ontogenesis. And that is just what an interpretative process involves: fitting the sign into a wider set of circumstances, a context. The external signals which the egg cell receives from other grasshop-pers affect the context in which the egg interprets the DNA.

When the locusts can no longer find enough food, the flock disperses, and the reduction in density induces a return to the original ontogenetic process.

The locust provides a dramatic illustration of a phenomenon that may well apply to many other life forms. Evolution has endowed the locust with a built-in interpretation mechanism. The locust species "has taught itself" to read the environmental conditions and report back on them (via a chemical signal) to its reproductive cells, its gametes, ensuring that they shape the members of the next generation to suit the prevailing conditions. When food becomes scarce, the switch is made to a new feeding strategy.

The other main sign operation involved in "evolutionary forgetfulness" is, as we saw earlier, reproduction. For simplicity's sake let us examine this process as it is ordered among species that reproduce by sexual means. What is happening here is that the conditions (in the widest sense of the word) for life, the ecological niche, are incorporated into the species or, rather, into the lineage (the species as an evolutionary unit) as the "rewriting of the species" described earlier is carried out. In every generation a meeting occurs between the species and its niche, and on each occasion this meeting will have a different outcome, one which manifests itself in the reproductive pattern. By reproductive pattern I mean the non-random[23] reciprocal selection of partners that determines which portion of the genetic material the lineage will pass on to the next generation. Through procreation, temporal and three-dimensional aberrations are translated into one-dimensional code; the (genetic) directory available to the next generation has been altered.

The nature of this process can be grasped only by viewing it as a *semiotic process*, i.e., a process of sign operation (fig. 7). Conditions for life are not automatically transformed into genetic material, and what does in fact take place cannot be defined as a straightforward mirror process. Every single generation is faced with a unique situation in which umpteen specific histories meet not only one another but also variable physical conditions, e.g., climatic variations. The lineage is not a sleeping partner, happy to take on the imprint of its surroundings. It makes more biological sense to sum up this phenomenon by saying that an active interpretation is effected. But, again, who is the "someone," the interpreter?

We have all been brought up in the Darwinist faith, so the obvious answer to this question does not readily spring to mind. The answer, however, is that the *lineage* interprets the niche conditions as a sign of the demands that will be made on future generations. Using the reproductive pattern as its tool the lineage alters the DNA text, which is then passed on. Thus it is the lineage rather than Nature as such that carries out the selective processes on which organic evolution is built. The term "natural selection" is in fact confusing and misleading.

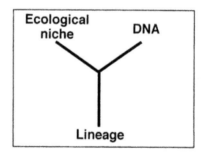

FIGURE 7. *Ecological niche—DNA—Lineage. Any given lineage interprets its ecological niche as a sign that the DNA pool must be modified.*

Doubtless for some time now objections to these statements have been hovering on the lips of the critical reader (she who is constantly challenging my intellectual equanimity). These objections will be much easier to answer once we have reached the end of this book. And I will return to them more than once. So I would ask you to be patient. But, at this point, let me just take issue with two of them. First, the one that says that the lineage is not an individual and that, consequently, it cannot be an interpretant. In answer to this I would like to point out that in 1992, as I was writing these words for the first time, the Danish people were coming to terms with the interpretation of the European Union's Maastricht treaty that had been made by Denmark in the referendum held the previous day. Denmark may not be an individual but that does not prevent us from saying that the result of the referendum constituted an interpretation.

Nor is the definition of an individual as obvious as one might think. Most people do, I am sure, consider themselves to be individuals. But as we shall see in the next chapter, one could very well argue that a human being is really the product of a collaboration between possibly hundreds of trillions of bacteria!

A more astute objection deals with the use of the interpretation metaphor. Obviously, if one insists that interpretation is a mental exercise of which only intelligent beings are capable, then of course neither an egg nor a lineage can interpret. But in that case I find myself stuck for a word to describe how the egg is in point of fact capable of transforming a one-dimensional DNA message into an exercise in time and space—a developmental process which, mark you, often goes wrong—in which case, the term misinterpretation is called for.

Ontogenesis is indeed a rather conservative process. Fortunately we can more or less rely on it—otherwise we would all vary much more in appearance than we do. But there are no grounds for regarding it as a predetermined process of the kind with which we are familiar from electric light or the movements of satellites. And as the American neurobiologist and Nobel prizewinner Gerald Edelman has shown in more complex organisms possessing a nervous system, we are dealing with processes that unquestionably cause anatomical variation.[24] "Genes cannot contain prior information about the spatial distribution of cells in the neural network," writes Edelman.[25] And he scathingly compares the idea of there being some kind of "electrician" within the genes, screwing the network together, with the belief of a bygone age that a little manikin, a homunculus, lay curled up inside the spermatozoa, needing only an egg cell to make it grow.

Let me sum up the conclusion which I believe I have now reached. The amazing ability to incorporate the present into the future which is the hallmark of all life is dependent upon its talent for forgetting, that is to say, for dying. Since organisms cannot survive in a physical sense, they are obliged to survive in a semiotic sense, i.e., by passing on coded versions of themselves or, as it were, signs. Heredity is semiotic survival.

Nevertheless, a tiny piece of the organism does also survive in physical form, namely the fortunate gametes that become the fertilized egg. A little egg such as this might be viewed as a template possessing, within its spatial structure, the directions necessary for interpreting the DNA inscription's mass of signs. In this way the code is translated back to the spatial structure.

Life is based entirely on semiosis,[26] on sign operations. Signs are flexible and open to misrepresentation, they reflect back on themselves and get shifted around and, little by little, they turn into new lumps in time and space, into habits. The living world is fraught with lies and deception, it can be awe-inspiring or deeply moving and, whatever else it may be, it concerns us. It is made of the same stuff as we ourselves are—it resembles us, because it dreamed us up.

In the next chapter we will learn in what sense this somewhat vague postulate might be said to be true; it being my plan, after all, to bring the natural world into closer contact with the human race than is usually the case. Usually biologists try to bring people close to Nature. I am going to adopt the opposite strategy, that of bringing Nature to people. What will come of this remains to be seen.

REPEATING   *3*

On Nature's tendency to acquire habits

Just imagine if someone were to switch off the Universe right now and start it up again from scratch with a new big bang. Would this result in my sitting here fifteen billion years later staring at a Macintosh computer screen while writing: Just imagine if someone were to . . . ?

The idea that everything in this life is predetermined by laws of nature—or, in olden times, by fate—to such a colossal extent that nothing in this world can occur by chance or because people decide it of their own free will—this theory is known as *determinism*. And it has been around for a long time. In his book on free will, the Danish author Villy Sørensen quotes the following fanatical observation by the Greek philosopher Chrysippus, who lived in the third century B.C.: "Why send for a doctor? If it is foreordained that I should recover then I shall recover without a doctor; if it is not foreordained that I should recover then I shall die despite the doctor. No, whatever you decide has already been foreordained."[27]

In the eighteenth century, "The Age of Reason," it is science that supplies the determinists with ammunition. "The keynote of mechanistic philosophy," writes Sørensen, "was not the realization that there is a natural explanation for everything. The Greeks had been well aware of that fact two thousand years earlier. No, it was the belief that every cause can have only one possible effect and that hence there is only one possible universal scheme which will inevitably run its course."[28]

In 1891 C. S. Peirce rounds fiercely on this "doctrine of necessity." He latches onto the odd fact that determinism refrains from providing an explanation for the most important question of all: where do they come from, these laws that direct the course of life with such fearful inevitability: "Law is *par excellence* the thing that wants a reason."[29] But the only possible way of accounting for the existence of natural laws is by assuming that they are the result of evolution. Admitting this, however, also means admitting that they cannot always have been absolute and allowing for the possibility that even today they might not always be followed to the letter. Thus, Peirce maintains, an element of indeterminism, spontaneity or absolute chance is introduced into the natural world.

Stretching the point a little, Peirce could be said, here, to be predicting the very results which, to almost everyone's surprise, came to light in physics circles in the 1970s in the form of the so-called chaos theory and "dissipative structures."[30] The fact that most physicists today would most probably agree that the universe could not be expected to follow exactly the same course after a new big bang as it did last time comes from learning more about the behavior of complex systems, as was illustrated by the "butterfly effect" mentioned in chapter 1. If infinitely miniscule fluctuations can have such dramatic effects, that must, surely, put paid to the determinist theory.

Or does it? The French mathematician René Thom challenges the notion that such fluctuations are prime events, believing as he does that they are simply manifestations of an underlying grand design. In France this has occasioned much highbrow debate which I will, however, refrain from pursuing further.[31] When all is said and done this is a metaphysical problem which can never be resolved conclusively. After all, how can anyone deny categorically that life is not merely the cunning execution of an endless sequence of steps in the great dance of the god Shiva?

It is in the nature of science—it is, so to speak, a matter of professional principle—to discover causes for the wonders of this world. Scientists have, quite rightly, done their utmost to extend the range of natural laws—and hence determinism—to bring as many as possible of this world's wonders under control, i.e., render them predictable. The marvelous thing about natural laws is, after all, that they make the world seem safe and predictable. For instance, it is only thanks to the Newtonian laws that we can be absolutely certain that the sun will come up tomorrow or that a paving stone is not going to leap up of its own volition and hit us in the face.

But useful as it may be to understand the principles behind natural phenomena, this does not mean that everything is bound by law. Yes, one might be tempted to look upon a belief in determinism as an expression of fundamental anxiety among people who dare not relinquish control. "Oddly enough, those philosophers who placed the strongest emphasis on the will also placed strongest emphasis on its constraints," writes Villy Sørensen in his book on free will.[32]

Like Peirce I prefer a philosophy which enables one to comprehend the world as a place where spontaneity is not rejected out of hand and where one can therefore entertain the thought that something radically new—i.e., essentially unpredictable—might be generated. A philosophy that has not already barricaded itself against the path to insight embodied by the question: Where did the natural laws come from? Because, quite honestly, why on Earth should they have been here all the time?

The crux of Peirce's metaphysics is that Nature has a tendency to "take habits." If Peirce is right, this would mean that it is not the laws of Nature which control the development of the cosmos; that the laws of Nature also had, at some point, to have originated—as slow-growing deterministic coral islands in a cosmic ocean of free-ranging vibrations.

But in that case we ought to be able to find some trace of such a "habituation" in the history of the cosmos. And in fact we can. From a certain angle, Peirce's theory—this tendency to take habits—appears to represent one of the poles in a continuous process of development, where the other pole could perhaps be termed "anarchy," Nature's tendency to reclaim its independence by means of new "inventions." For simplicity's sake let us call these two opposing forces in natural history *fate* and *freedom*—though without attributing any more to these two terms than we have already done.

But freedom or the lack of same are dialectic quantities, since, paradoxically, at one particular level a lack of freedom can actually pave the way to a different sort of freedom at a higher level. Thus in retrospect we can see that only when lack of freedom had reached a level at which the world became, to some extent, predictable—i.e., when certain habits or laws had become firmly established—did it become possible to develop the actual *ability to predict*. As we saw in the previous chapter this ability is the very hallmark of all life-forms. In a world where nothing was predictable, Life would be out of a job. And the wealth of inherited experience lodged inside the genetic material of an organism would be totally useless without the possibility of cherishing reasonable expectations of the future.

Let us now examine in more depth the two terms *freedom* and *lack of freedom*, as they pertain to the way in which matter and energy are arranged within a living cell. In fact the words almost make my point for me. Usually, when we employ the word *cell* in a human context we are referring to what we call the loss of liberty. As prisoners we are deprived of our freedom, inasmuch as we cannot leave the prison cell whenever we feel like it. Similarly it can be said that atoms caught by a cell and absorbed into its structure are deprived of their freedom. Until then a carbon atom might, for instance, dream of seeping down, as bicarbonate, into the ground water, whence it would flow out to sea and travel round the world. But once it has been sucked up by the roots of an oak tree and captured by a cell at the growth point of the trunk, the atom runs the risk of having to wait a thousand years for the oak tree to be felled in a storm and rot away. Only then will the carbon atom be set free. By being captured by life the atom loses its freedom.

I hope my readers will forgive this rather frivolous method of explaining things—far be it from me to credit atoms with traits such as dreaming. But it may perhaps be easier to see what I am getting at if one puts oneself—in this case—in the atom's place.

My point is that, as an earthly phenomenon, Life itself exemplifies Nature's tendency to acquire habits. With the emergence of an arrangement of matter and energy as unique as that found in a living cell, so too a new and intricate pattern was established in the world—a pattern that could be repeated *ad infinitum*. And repetition is of course the epitome of habituation: the key to predictability, law, and order. Again and again water and carbon dioxide now had to see themselves being coupled together to produce carbohydrate in a cell before being broken down once more into water and carbon dioxide—over and over again.

How life first came about is something at which we can only hazard a guess, but there can be little doubt that it did. Somehow or other inorganic matter managed to form itself into an ingenious system, the cell, which imposes constraints on its constituent parts, its atoms or molecules, inasmuch as it is the cell as a whole that sets the limits for what the individual molecule may or may not do.

The American evolutionary biologists Niles Eldredge and Stanley Salthe have used two terms to describe this situation: *initial conditions* and *boundary conditions*.[33] Initial conditions are the physical and chemical characteristics associated with the mass of molecules which make up the cell. Boundary conditions are the historic pattern imprinted upon the cell at the very moment of its creation through cell division, i.e., the cell's internal organization. The old idea of a cell being like a sack full of proteins and all sorts of other good things has been supplanted by the contemporary view of the cell as having a complex inner structure that bears more resemblance to the structure of a city than to the structure of a sack of flour (more on this in chapter 6).

But the point at which the true focus of this account starts to become clear is when we discover that it is precisely this freezing of the cell's chemical make-up which institutes a totally new kind of freedom, one which I will call *semiotic freedom*.[34] Because even the single-celled organism knew a little trick which proved most effective in tempering the growth of predictability. It was able to describe itself—or at least key aspects of itself—in an abstract code embedded in the string of DNA molecule bases. Fragments of this coded self-description could then be copied, sometimes wrongly, and traded with other members of the same species—or even, on occasion, with members of other species.[35] The never-ending sequence of "mistakes" and "misunderstandings" that put all life-forms on Earth into a constant state of flux, the sequence which we call organic evolution, was set in motion.

The predictability of chemical laws facilitated the establishment of unpredictability at a biological level. The tendency to acquire habits—fate—had been overcome, for the time being at least.

But fate was not about to give up. Slowly and steadily it proceeded with the job of establishing higher forms of predictability, new habits.

The first breakthrough in this endeavor came with the establishment of the *eukaryotic cell*, a cell type which forms the foundation stone of all "higher" life-forms. Compared to the primitive *prokaryotic cells,* which resembled the bacteria of our own day, eukaryotic cells are both very large and very complex. First and foremost the eukaryotic cells possess a unique internal structure, the nucleus, repository of the genetic material. But the eukaryotic cells also contain a large number of other, lesser "bodies," so-called organelles. Of these, probably the best known are the *mitochondria* which are charged with the important task of producing energy for the cell. In plants, photosynthesis is managed in similar fashion by organelles known as *chloroplasts.*

As far back as 1893, the German biologist A. Schimper outlined the theory that the chloroplasts in plants were derived from cyano-bacteria (perhaps better known as blue-green algae). Later, and quite independently of one another, a Russian and an American biologist both advanced the same theory. But it is only within the last two decades that this theory has, at long last, won general acceptance. However strange it may sound, it seems likely that the modern-day eukaryotic cell was generated by some kind of symbiosis, whereby a great many tiny prokaryotic cells combined to form one large departmentalized cell, the eukaryotic cell.

One reason for the acceptance of this so-called endosymbiosis theory is that numerous examples of this kind of symbiosis are now recognized. One of many intriguing instances centers around the termite. As we know, termites eat wood and are notorious for their ability to reduce a wooden house literally to sawdust in next to no time. But the termites themselves cannot digest the wood's ground substance, cellulose. Like ruminants, they have to depend on an intestinal flora of microbes breaking down the cellulose for them. A small single-celled eukaryote organism, i.e., a protist by the name of *Mixotricha paradoxa,* has a vital part to play in this digestive process. And in return for breaking down the cellulose, Mixotricha is provided with a constant supply of ready-chewed wood.

But Mixotricha is an odd creature. For, though we may call it single-celled, it moves by dint of five hundred thousand miniscule bacteria, the *spirochaetas,* which cling to the surface of the eukaryote cells. These spirochaetas are closely related to the syphilis bacteria and, like it, they have a little whip (a flagellum) at one end which they can rotate and, thus, propel themselves. And, fitted with five hundred thousand spirochaetas, it has to be said that *Mixotricha* certainly does get around. But its refinements do not end there, because it has been proved that *Mixotricha* has no mitochondria; in other words it lacks what might be called its inner power plant. So instead *Mixotricha* has formed an alliance with a kind of bacteria living *inside* the cell, where they are apparently happy to carry out the job of producing energy for the protist in exchange for being sure of a steady supply of food.

The question is: is *Mixotricha paradoxa* a single-celled organism or is it, rather, a colony of more than five hundred thousand bacteria representing several different species? In other words, is *Mixotricha* one individual or a large number of individuals? From a biological point of view this situation would probably best be described as an extremely close symbiosis, an *endosymbiosis*. Translated literally from the Greek, symbiosis means the state of living together, and in biology it denotes a very close and enduring association between two species. Most gardeners are familiar with the symbiosis that exists between ants and aphids, in which the ant protects the aphid and, in return, the aphid allows itself to be milked of sucrose. In this case the two parties involved are doing each other a good turn, but symbiosis can also be malignant, as in the case of parasites. With *Mixotricha* too we are dealing with endosymbiosis since, here, some of the bacteria are actually living within the cell, where they take the place of the mitochondria.

But where does endosymbiosis end and individuality begin? What we, today, can discern in *Mixotricha* is presumably exactly what took place one-and-a-half billion years ago when the very first eukaryote cell was created. Through time the original resident bacteria simply lost a considerable proportion of their independence and became the mitochondria of our own time. To this day the mitochondria carry traces of their bacterial past, in that they have their own DNA, which bears a closer resemblance to bacterial DNA than it does to that of a cell-nucleus. And Lynn Margulis, the biologist who has fought harder than anyone for the endosymbiosis theory, even claims that not only the mitochondria but also other primary cell functions such as cell division and cell mobility are controlled by organelles of bacterial origin.[36] In a sense a human being too is basically an endosymbiotic system comprised of hundreds of trillions of bacteria. Viewed in that light bacteria are the only true individuals in this world, all other life-forms being mere combinations of bacteria!

With the formation of the eukaryotic cell the individual prokaryotic cells were to some extent deprived of their freedom, subjected as they were to the conditions set by the eukaryotic cell. Their own particular vital processes had to be coordinated and become more specialized in deference to their collective fate. A new habit had emerged. But once again this renunciation of freedom was to spark off an entirely new form of creativity which was simply acted out at a higher level of complexity.

The eukaryotic cell did have certain talents which far surpassed those of the prokaryotic cells. This becomes most evident if we compare the communicatory skills of the two cell types. The bacteria (prokaryotes) are indeed busily engaged in the comprehensive exchange of signs—in the guise of DNA fragments—which can, for example, be transmitted by means of particular bacterial viruses.[37] On the other hand, the number of other sign operations taking place between them is very limited.

In the eukaryotic organisms, however, DNA communication is a family affair. DNA is transmitted almost exclusively to the next generation. There is a tradition in biology for depicting time as a vertical axis (cf. the word *descent*), and it can therefore be said that the eukaryotic organism translates genetic communication into a purely vertical phenomenon, transmitting from parents to progeny—in other words, a *vertical semiosis*. To counteract this "privatization" of the genetic material the eukaryotic cells have, however, developed ingenious and efficient methods of communicating with one another by chemical means, primarily through physical contact. Special proteins on the surface of the eukaryotic cells can, so to speak, poke their noses into their neighbors' affairs (something which will be discussed in greater detail in chapter 6). In other words, a form of communication evolves that is not based on signs in the form of genes but on signs in the form of proteins or other types of chemical compound. One could call it *horizontal semiosis*, the exchange of signs through the three dimensions of space rather than through time. Not so much genealogical semiosis as ecological semiosis.[38]

Even prokaryotic cells had the ability to link up into chains or clusters, but in the eukaryotic organisms this multicellularity was combined with specialization among the cells, whereby the cells somehow learned to communicate with one another on the delegation of work: which would attend to the production of gametes, which would lash the flagella, which would pick up signals from the surrounding world, etc. The eukaryotic world's multiplicity of cell types was the opposite of the prokaryotic world's multiplicity of DNA fragments.

Once the cells had surrendered their anarchic autonomy and submitted themselves to the greater whole which we call the organism, the stage was set for the creation of the most fantastic life-forms. Now the pace hotted up in the development of sophisticated sensory apparatus and corresponding nervous systems which would enable animals to form fine-tuned internal impressions of what lay round about them, their surroundings. Their subjective experience of the world, their *umwelt*,[39] became fraught with detail and the horizontal semiosis—that process of lies and deceit, of play and sexuality and Heaven knows what else that binds the creatures of the Earth to one another—grew and grew in abundance. Once again freedom had drawn the long straw.

But Fate had a joker up its sleeve—and one which looked as if it could cause trouble. While there was no way of quashing semiotic freedom, it could be stamped with patterns that would ensure some form of order. And what was this regulator, the joker in the pack? The ecosystem. As the network of food chains gradually began to seal itself off with a series of closely integrated circuits, a higher form of logic emerged, one which seemed to compel individual species to fulfil particular roles. Nature had arrived at a situation which ecologist G. E. Hutchinson so appositely dubbed "the evolutionary drama in the ecological theatre." Granted, new species continued to appear, and marsupials and mammals replaced the reptiles but, as the following illustration (fig. 8) shows, any originality was in a sense illusory, bound as it was by the predominant pattern of ecological niches which this planet's ecosystems happened to have on offer.

So had Fate and habituation finally won the contest? Was freedom but an illusion? We will never know. But freedom did have yet another card to play—and as yet that card does not seem to have been covered.

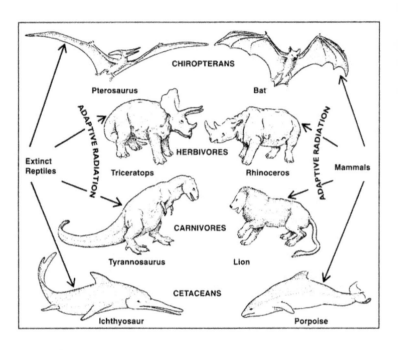

FIGURE 8. *This illustration shows how mammalian evolution led to life-forms closely resembling those reptiles that had occupied similar niches in Nature's household at an earlier stage. Taken from Wilson et al.* [40]

Among all the roles in the ecological theatre there was one pertaining to creatures with lengthy life histories and an especially well-developed talent for capitalizing on their experience. Often these creatures, the apes, had developed brains capable of accommodating an extremely complex image of their surroundings, a very sophisticated umwelt. And among these creatures was one in particular that played its role so uncommonly well that the role became reality in all of its dreadful isolation.

There came a day when this creature realized that it was itself an *umwelt* builder; that its role was, in fact, a role; that other creatures performed other roles and had different kinds of *umwelt*; that the world was one thing and *umwelt* another; and that, when one died, this *umwelt* would actually disappear while the world as such would carry on. In short, this creature perceived its own existential alienation from the world.

Fortunately, before our creature made this dreadful discovery—

which, had it been totally unprotected, would have driven it insane—it had succeeded in safeguarding itself through the development of a gift for empathizing with other similar creatures. (We will return to this point in chapters 8 and 10.) This empathic gift enabled the creature to create a common bond of a quite unprecedented nature: a double bond founded on the need to share the *umwelt* with one another, i.e., making private experiences public property, turning the subjective into the objective. To cut a long story short this creature, with whom the reader is already identifying strongly, invented the spoken word.

As opposed to the primal world in which the laws of Nature have long since established a firm foothold, the world of language is free; in that world anything can happen. In this extraordinary manner—through the creation of a linguistic bond which some refer to as the intellect—humanity freed itself from Fate.

But we had better take care. Nature's penchant for forming habits does not stop at language. Fate still has a few tricks up its sleeve, some of which are very much geared toward standardization. Religion is probably the most intransigent in this respect but there are times when politics can seriously stifle the imagination. And yet nothing has so far been able to suppress the fundamental freedom of the intellect for long. The anarchic nature of human thought and imagination appears to defy any and all civilizing influences.

But we will have to close this account of the battle between Fate and Freedom by acknowledging that Fate can afford to bide its time. Its timescale is measured not in thousands but in millions of years. So who can say that it is not sitting there thinking, "Ah, let the child have its fun."

Peirce's theory that Nature has a tendency to take habits does in fact make good sense. Just to recap:

The Earth's currents of matter and energy were gradually channeled through the increasingly complex recesses of all its many life-forms. Physico-chemical habits became biological habits. Primitive cells were organized into endosymbiotic patterns which we call eukaryotic cells. Eukaryotic cells acquired the habit of working together as multicellular organisms which in the course of time adapted to the prevailing logic of the ecosystems. The stabilization of living conditions under this form of logic made both longevity and intelligence an advantage and, hence, the logic of the ecosystems was eventually shattered by the appearance on the scene of humanity with its formidable talent for bossing pre-human life around. But even human beings could not shrug off this knack of forming habits. Each civilization is a manifestation of the way in which a new master plan is accepted, a plan that will significantly boost or diminish the unpredictability of human thought and deed.

Thus the history of the earth chronicles the gradual formation of an ever-widening range of stabilizing factors. While the original prokary-otic cells could only determine the conditions for an insignificant proportion of the Earth's currents of matter and energy, we are now well on the way toward a situation in which many of these currents are determined on a global scale. So stable is the plethora of gases in the earth's atmosphere that we can only assume they are being subjected to very strong feedback control from the Earth's collective network of ecosystems, the biosphere.[41] But this stability is only tenable if we human beings establish the cultural conditions for an end to its erosion.

Descriptions of the dynamic force presented here often employ the term *emergence*, a word meaning "birth" or "manifestation" used to describe the creation of an "entity" with qualities of which its compo-nent parts give no hint.

A classic example is the quality of "being wet." Hydrogen and oxygen are both gases. Combine these two, however, and something wet—water—is produced. The wetness is said to be an emergent quality which could not be deduced from a familiarity with oxygen and hydrogen. But I have always felt that this was a rather stupid example. Wetness is a sensation and as such there is obviously no way that it can be present in oxygen or hydrogen. And yet it is not so hard to identify, in advance, water's knack—*when people experience it*—of manifesting itself as wetness.

Emergence is, nonetheless, an excellent word to describe historical development or evolution.[42] What we can see in the history of the natural world is the successive formation of new and well-organized entities or systems that impose a set of boundary conditions or constraints on their individual components—or sub-systems. With the advent of multicellular life the individual cell not only lost its freedom of movement, it was also assigned a particular specialization—to be, for instance, a liver cell, a nerve cell, or a skin cell—and was thereby definitively cut off from the rest of its inner potential. Multicellularity is therefore an emergent quality in the sense that it establishes a set of hitherto nonexistent rules.

The ecosystem too is an emergent quality. Choice in the ecosystem is restricted to certain specific ecological niches and those alone. So there is no way that one can keep track of the ongoing development of individual species without considering the new logic now prevailing over the system as a whole. New species come and go but the ecological conditions remain the same, with the result that, at the species level, evolution has largely become a repetition of the same roles with a new "cast."

There is nothing mysterious about emergence. In the case of *Mixotricha,* the termite's intestinal microbe, for instance, we were presented with a contemporary endosymbiosis similar to that which, in all probability, once led to the emergence of the eukaryotic cell as a superior organism. The only mysterious thing about the term *emergence* is, if you like, that it forces us to accept the fact that Nature is comprised of a hierarchical system in which, at every level, we are faced with a set of boundary conditions which must be acknowledged as independent factors in the evolutionary process. This is an acknowledgment which, it must be said, is never very popular in traditional biological circles since it could easily end up destroying the mathematical statutes they are so keen to discover in the natural world.

Galileo's time-honored assertion that the great Book of Nature is written in the language of mathematics remains the credo of the scientific world—its article of faith. And yet, as things stand, there is no evidence to support such a belief. And we could be excused for wondering at the totally irrational way in which scientists hail mathematics as Nature's guiding principle.

At times it actually seems as though they would rather produce incorrect explanations than discard their mathematical explanations. Thus, one of the assumptions made in many genetically based evolutionary models is that reproduction among test subjects is conducted through the quite random selection of partners; that is to say that large individuals do not necessarily mate with other large individuals and so on. When, in actual fact, we have long known, as Richard Lewontin puts it, that ". . . organisms virtually *never* mate at random. They mate assortively by color, size, location, age, activity. . . ."[43] Add this exasperating state of affairs to the reckoning, and mathematical models can easily become very complex and prediction much less unequivocal.

I am somewhat skeptical of this worship of the God of Mathematics. I have a suspicion that, deep down, Galileo's credo is an expression of human reason's wish that the world should always resemble reason itself. But what if it does not? What if the world bears more resemblance to some crazy story, or a fairy tale? Might mathematics not conceivably be an unnecessary detour on the road to understanding it, rather than a shortcut?

Should we not at least keep open the option that the world is in the purest sense a creative place, the future of which, by that very token, is impossible to predict (after all, being predictable would mean being void of creativity)? And, that being the case, emergence cannot be mathematically pliable, since that would imply that the same initial conditions could lead to different endings in different drives within the computer. This would not prevent mathematical models from remaining indispensable tools in limited areas of real life. But it would remove all grounds for the arrogance with which Science often wriggles out of taking the incalculable aspects of our world seriously.

Or perhaps I am mistaken. Perhaps one really could create a sort of mathematics of unpredictability. But at this point inspiration fails me, so I will confine myself to saying: in that case, that's fine by me. What I object to is the desire to force life processes into some kind of mathematical equation, with a certain quantity set on either side of an equal sign. Because this implies that the quantity on the right-hand side is "no more than" the quantity on the left-hand side, a statement which can only be true if creativity has been weeded out of the system, if it has in other words been killed off.

To my mind, if the existence of emergence means that current mathematical models lead to misleading descriptions, then the solution must lie in finding other means. I nominate semiotics, the study of sign games.

And sign games really start to become fun just at the point where mathematical difficulties begin to arise: namely, where self-reference crops up in a system. And that is what we are now going to take a look at.

On life and self-reference, on subjectivity

The following sentence means exactly what it says: *This sentence contains three rong words.* Now that cannot be right, can it? After all there is only one wrong word—the word "rong." In order to read correctly this sentence would have to be altered to: This sentence contains one rong word. But the original sentence did actually contain three wrong words: "Three" should have been "one," "rong" should have been "wrong," and "words" should have been "word." So the sentence was true after all, wasn't it? And yet as soon as we say that, it is wrong again. Because then again there is only one wrong word. In other words, when the sentence is true it is false, and when it is false it is true.

Here we are confronted with a specific type of paradox, one which has had a crucial part to play in twentieth-century philosophy. This paradox is known as "the liar paradox," and in its simplest version it goes like this: "I am lying." And here the reader is obliged to assume that my assertion that I am lying is a lie. Or, to put it another way, that I am speaking the truth—which immediately implies that I *am* lying after all.

In actual fact there is no limit to the potential complexity of such paradoxes. In all probability they crop up all the time in conversation without our ever being aware of it. Paradoxes can in fact be constructed out of several sentences which are not, in themselves, problematic. As, for instance, here: "The following sentence is false. The preceding sentence is true."

In all of these instances the self-same problem arises, that of *self-reference*. The sentences refer to themselves in such a way that they contradict themselves. Self-reference is not necessarily paradoxical but it is the bane of logic because it breaks the rules, so to speak. It puts us into a kind of hall of mirrors where it becomes impossible to distinguish between a statement and what that statement concerns. Like a map which is so detailed that the cartographer and the map he is making are swept up into it. Or like the first sentence of chapter 4.

The more mature American reader is bound to remember the little girl on the Morton Salt container—"when it rains it pours." She holds an umbrella and pours salt from a Morton Salt container on which she appears, holding her umbrella and pouring salt from a Morton Salt container. . . . (The effect is muted slightly in the current streamlined design in which only the shape and color of the container identifies the object she is holding.) For an entire generation of children, she represented their first encounter with the bewildering depths of self-reference. I must have been about the same age when it dawned on me, on those occasions when I found myself being sucked down that angst-ridden elevator shaft in a flash of terrible doubt as to whether I really existed, that I was in very much the same situation as the Morton Salt girl. If I could feel doubt then I must exist. And in order to conquer my doubts all I had to do was doubt![44]

Douglas Hofstadter dubbed such chains of self-referential elements "strange loops."[45] And in this computer age it is a safe bet that every older child is familiar with all sorts of sophisticated versions of this phenomenon.

But in the early years of the twentieth century these paradoxes were giving mathematicians and logicians gray hair and the mysteries of self-reference were considered to be the very root of all evil. One approach involved simply forbidding self-referential statements on the grounds that they were meaningless. This approach had been introduced by Bertrand Russell and Alfred N. Whitehead in their seminal work *Principia Mathematica* (1910–1913). Such a prohibition might work for a while in mathematics and in logic, but it would never do for everyday conversation—since it would be virtually impossible to carry on a meaningful conversation without using the word "I" and thus instantly violating the prohibition.

But the Russell-Whitehead solution finally collapsed in 1931 when the Austrian philosopher Kurt Gödel demonstrated that every mathematical system—*Principia Mathematica* included—involves true statements that cannot be proved within the system without this giving rise to contradictions. Gödel introduced a special code, the gist of which was that statements about numbers can themselves be expressed by a number. This made it possible to create a self-referential sentence which is also correctly formulated within the *Principia Mathematica* system. Hofstadter elegantly sums up Gödel's discovery in the following colloquial sentence:[46] "This statement of number theory does not have any proof in the system of *Principia Mathematica*." This is quite clearly a self-referential sentence and one which has very craftily wormed its way to the very heart of what was at that time the most carefully nurtured of all logical and mathematical castles in the air.

In 1934 a Polish logician, Alfred Tarski, devised a theory closely akin to Gödel's, namely, that the only meaningful way of employing the term "truth" is in a special *metalanguage*. The use of the prefix *meta* has now become so much a part of modern thought that we have no choice but to confront it.

Example: I meet a man who says: "The moon is a green cheese." Tarski would classify such a sentence as "object language," that is to say a language which refers to concrete objects such as moons and cheeses. Having listened to this man I then feel like writing: "That man's statement was untrue." This sentence belongs to a metalanguage, since it is a sentence that does not refer to objects within the world but to another sentence. It exists on what is often referred to as a *metalevel*. And by writing it down I have now employed a metametalanguage, and so on. According to Tarski we can only express the truth in the following metalanguage: the sentence "The moon is a green cheese" is true if, and only if, the moon actually is made of green cheese.[47]

To philosophers and logicians like Tarski such precise specifications are of course vitally important. But in practice we cannot combat paradoxes in this manner. In everyday language—as well as in serious books such as this—we inevitably intertwine the object level, the metalevel and the metametalevel in the most baffling fashion. Otherwise we would never be able to tackle the complexities of real life.

Instead of being afraid of paradoxes, instead of forbidding us to be self-referential or forcing us to observe various rules and regulations as to what may be said at which linguistic levels, I believe we should delight in them. Who knows, maybe these paradoxes spring from the self-same source as the creative spirit, namely the fact that it is not just a matter of talking or thinking, but that "someone" talks or thinks? The idea that it should be possible to devise a formal system of rules by means of which the world will graciously allow itself to be bound and gagged comes from the same stable as the determinist theory. That these theories boost Reason's self-confidence is no good reason for believing in them. And it is hard to reconcile them with the lessons we learn from science and from our everyday lives.

When all is said and done there is something absurd about believing that we can escape the snares set by self-reference. We are, after all, creatures of flesh and blood whose thoughts and speech have in every respect grown from our actual—temporal and spatial—life stories. By this I am not trying to say that we all necessarily "talk through a hole in our heads," but on the other hand I think we would do well to admit that we could not talk without a hole in our heads. We will be examining the physical aspects of language in more depth in chapter 8. But now I want to return to my real reason for bringing up the subject of the self-reference maze: the fact that, at heart, all life is founded on self-reference.

Every single organism on this Earth contains some sort of self-description in DNA form—something which we have already covered (in chapter 2). And now we must study the essence of this self-description in greater detail.

Earlier in this book I compared the DNA code to a collection of recipes in a cookbook. But it would in a sense be more correct to compare it to the score of some vast choral work. Because the development of the embryo is in fact accompanied by a simultaneous reading, carried out by a multitude of "voices" in the form of genes. And it is the coordination of all these readings, through which the choral work manifests itself as a unified whole, that lends grandeur to the interpretation.

Where the development of the embryo is concerned this coordination appears to be dependent upon a finely tuned collaboration between the individual tissue types, each of which is highly self-regulating. In short, the conductor does not seem to have any counterpart in the ontogenetic process, where the individual "singers" and "musicians" regulate their growth as a group, through a system of reciprocal communication (endosemiosis) about which we still have only the vaguest inkling. At any rate, the genome (the genetic material as a whole) cannot in all fairness be compared to the conductor, because the genome is only the score. Not that *that* is anything to snap our fingers at either—just think of the Saint Matthew Passion.

One of the crucial features of the coded DNA molecule version of the organism—and one which also typifies both the score and the text—is that it is *digital*. The word *digital* comes from the Latin word *digitus* meaning finger. And the main point about digital codes is that they are disjointed, much in the same way as the five fingers on a hand, which are separated from one another by spaces just like the numbers 1–2–3–4–5; or like words, which are of course comprehended or written down one at a time; or, for that matter, like musical notes. Not all codes are digital; in fact by far the majority are analogs, i.e., they are based on some kind of similarity (analogy) with whatever their code represents. Probably the best-known examples of these two types of

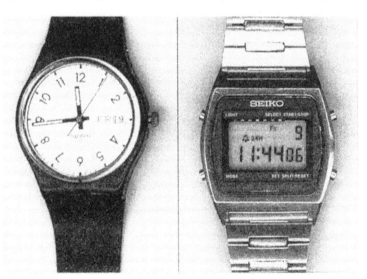

FIGURE 9. *Examples of analogic and digital codes.*

code are presented by watches (fig. 9). On an "old-fashioned" watch the small hand's journey around the watch face is an analog for the sun's apparent journey around the Earth. On a digital watch the movement of the sun is represented instead by a series of disjointed points each of which is defined as a combination of figures indicating hours, minutes, and seconds.

In using the terms self-reference or self-description of the digital DNA version of the organism I have, in fact, raised a very weighty question: What do we mean by the word "self"? And when is something a "self"?

Throughout this century the question of how life originated has been a bone of contention in biological circles. One school of thought maintains that the nucleic acids (DNA or RNA) came first; another school claims that the first thing to emerge was a proton cell packed with proteins and that DNA and RNA were later "inventions."[48] This dissent can be symbolized by the question: Which came first, the chicken or the egg? There is a theory, in vogue at the moment, which says that RNA (the egg, that is) was the moving force behind the beginnings of life.[49] But if the whole thing started with RNA/DNA ought we not to talk about the organism as being a coded version of the DNA? Would the organism not then be regarded as the DNA analog's self-description rather than the DNA's being the organism's digital self-description?

In the light of what we now know neither of these two possibilities can be ruled out. But no matter whether it was the cell (and the cytoplasm) or the RNA that came first, the way I see it we would have to say that a "self"—and hence life—does not exist until both of these versions are present. For the DNA "message" would be null and void if it were only a copy of itself, and the organism's message would be mute if the DNA could not ensure its dissemination. Kantians may take issue with this distortion of the master's celebrated observation on categories and sense perception. But my intuition tells me that this parallel is not so wide of the mark.

Life is dependent on this semiotic interplay between the analogic and digital versions of the message—in other words, on *code duality*—and "the self" can only refer to the union of these two versions within an organism. Just as the self in a human being involves both the physical and linguistic elements, "the biological self" involves both the cytoplasm and the DNA.

And just as human beings use words, the digital code of language, to reflect on themselves and describe themselves, organisms use the DNA's digital code to create their own form of self-description. In both cases this amounts to a fragmented and inadequate translation which, if it is to mean anything, must be reconstructed by a person or an egg as the case may be. Errors both in the self-description and the interpretation ensure that the code duality is never allowed to lapse into a state of conservative stagnation but that it will keep on generating new grist for the evolutionary mill.

But I am on dangerous ground here. Because, while scientists may be averse to the idea that the natural world is populated by subjective and hence fundamentally unpredictable beings, humanists and theologists are just as averse to the concept of subjectivity being sullied by application to earthworms and seaweed. So it is essential that we choose our words carefully. I will shortly be returning to this point, but before that I must just elaborate a little further on the term *code duality*.

The reader may well find the thought of organisms as analogic codes confusing. Codes for what, one might ask. Let us be more specific. Every single crocodile embodies both the essence of being a crocodile, "crocodileness" (the message handed down to it through the genetic material), and the elements that make it one particular crocodile. The second message is a kind of meta-message supervenient to the bloodline's digital message. The crocodile is an analogic code in the sense that it enters, among other things, into a mating semiosis which, in principle, involves a good many crocodiles (through competition, etc.). Ostensibly, the message is transmitted by the fertilized egg cell the crocodile once was, but it also involves the egg cell's spatial interpretation of another message, the digitally coded message that, at one time, lay tucked away inside the crocodile egg's own genome. And, as the mating semiosis runs its course, this message is received—and interpreted—by other members of the same species. Generally speaking an organism conveys a message about its evolutionary experience. But for whom is this message intended? Here I have to admit I am at a loss. Parts of the message—such as smell, sound, behavior, appearance—are of course intercepted by other members of the same species; these same parts, or others, are intercepted by individuals from other species in the ecosystem. And of course the biologist eventually intercepts a good deal of the message.

But in general there is no way of telling what the purpose is of all the communication taking place on our planet. Take people, for example. For the most part they communicate because they enjoy doing so, rather than because they have something special to impart. And as far as one can gather, the lion's share of messages transmitted on this Earth are sent without any thought for the actual recipient. Think, for instance, of the faces we pull while watching a film or "thinking out loud," or the signals transmitted by the way we hold ourselves, or the way we dress. So perhaps it ought to be said that all organisms participate in life as bearers of an evolutionary message aimed at the new creations of a—to some extent—unpredictable future.

With this in mind one could perhaps question the point of talking in terms of codes and messages. To this I would say, first of all, that it seems more appropriate and more satisfactory to speak of living creatures as messages rather than as vehicles for survival.[50] And secondly it is also more logical. There are two distinct reasons for this.

The first is that most of what happens between animals themselves or between animals and their environment is triggered by or carried by stimuli which, from a physico-chemical point of view, are negligible when set against the volume of matter and energy thus activated. To take an example: a dog weighing one hundred pounds suddenly dashes off at top speed. Now obviously there is a physical and chemical explanation for its taking off like this. But while it may be possible to account for the dog's muscle contractions biochemically, that still does not truly explain this phenomenon. It would be truer to say that the dog has received a message, namely the scent of another dog. The other dog is, in fact, a message.

Messages can initiate events which have no rational connection with the physical and chemical elements of which the message is comprised. As, for example, with the single shot fired in Sarajevo in 1914 which triggered off a world war.

And at another level: The swallow which, in its graceful flight, manages almost miraculously to avoid all obstacles in its path, is a product of history with everything about its body and its behavior representing millions of years of evolutionary experience in the physics of flying and the biology of the insect world. Is that not a good enough reason for calling the swallow a message? Goethe wrote:

> *War' nicht das Auge sonnenhaft*
> *Wie könnten wir das Licht erblicken?*
> *(If the eye were not sun-like*
> *It could never behold the sun.)*

To which Jakob von Uexküll added:[51]

> *If the flower were not bee-like*
> *And the bee were not flower-like*
> *The unison could never be successful.*

Which brings us to the second reason for describing organisms as messages. Only by regarding organisms as bearers of evolutionary messages can we apprehend the unique aspect of all living creatures which I will call *intentionality*. This term, *intentionality*, has given philosophers many a headache.[52] In the phenomenological philosophy of Brentano and Husserl intentionality is connected to the idea that our mental states are always "about" something "out there." This inescapable "aboutness" seems to be a broader phenomenon than just consciousness. From a biological point of view there is nothing surprising in this "aboutness." Only animals have nervous systems and brains; these have never been found in plants—and from the dawn of evolution their purpose has been to control bodily actions, behavior.

It is a well-known fact that animals can and do dream. This implies that mental states can be uncoupled from bodily actions. But the extent to which behavior and the mental activity that characterizes the human mind can be uncoupled is probably unique to that specific animal. This uncoupling causes philosophers to wonder how it can be that mental states are always "about" something. But this is because they never consider that mental "aboutness"—human intentionality—grew out of a bodily "aboutness" (i.e., the behavior necessary for assuring reproduction and survival)—what could be described as *evolutionary intentionality*, the anticipatory power inherent in all living systems. We still cannot escape the fact that our minds remain embodied.[53]

To say that living creatures harbor intentions is tantamount to saying

INVENTING

47

that they can differentiate between phenomena in their surroundings and react to them selectively, as though some were better than others. Even an amoeba is capable of choosing to move in one direction rather than another. It will, for example, generally gravitate toward the richest source of nourishment. And although there is a purely practical, biochemical, explanation for this faculty the true explanation must perforce be of a historical nature, since it has to be able to account for how, in evolutionary terms, such a faculty has originated.

Which brings us back to the unique mode of survival which is open to living creatures by reason of their mortality. We say that they survive semiotically inasmuch as they bequeath self-referential DNA messages to the next generation. Thus we can see that, in biology too, self-reference engenders infinite depths in the form of evolution's genetic hall of mirrors.

Intentions presuppose temporality. If the present second were to last forever all of our intentions would be to no purpose. Intentions are dependent upon being able to anticipate the future. And the future can only be anticipated on the basis of past experience. A self-referential system based on code duality will either die out or it will begin to invent a history, i.e., it will develop into an intentional system. And this derives from the unique set of dynamics inherent in the interplay between the digital code and the analogic code.

The digital version of the genetic message, the DNA phase, is a passive version—in exactly the same way as the cookbook is passive. Actually the DNA message is more or less *shielded* from all the ups and downs of life. The idea is not, as the French biologist Jean Baptiste Lamarck (1744–1829) believed,[54] that the life of the organism should leave a clear imprint of itself in its self-description, its genes. According to Lamarck, wader birds had long slender legs because for generations they had been doing all they could not to get wet. Thanks to their persistence the legs of each generation of waders would become just that little bit longer—an elongation from which, Lamarck has it, their offspring then benefited. But no one has ever been able to prove this theory—that "acquired characteristics are inherited"—and as far as we can tell it is quite fallacious. It is, however, worth noting that even though acquired characteristics are not normally inherited at the individual level, this kind of inheriting is exactly what does take place at the population or species level. In the case of the population, the whole point is that the gene pool of each generation is the product of what happened to the previous generation. Lamarck's "crime" was, therefore, that he did not differentiate between the individual and the population. Yet this differentiation happened to be the self-same unprecedented and momentous innovation that Darwin had introduced. And, almost a full century after Lamarck's original theory was advanced in 1800, it was the Darwinists who developed the idea that the key to evolution should be sought at the genetic level. Seen in that light, Lamarck's theory of evolution naturally collapses. On the other hand it has to be said that the problems which Lamarck was attempting to solve have more or less faded into obscurity under the myopic lens of neo-Darwinism. Energetic attempts are now being made to develop a more broadly based theory of evolution, but these are, sadly, meeting with widespread theoretical apathy, not to say open hostility.[55]

Code duality provides us with a key to understanding why DNA has to be shielded from life, because one of the important benefits of the digital code is that it can stand the test of time. Plato may have died several thousand years ago but his written works are still in existence today. Digital codes ensure some continuity between the past and the future; they are codes for memory. But codes for memory must be immune to any fleeting transient variation. Acquired characteristics cannot be inherited, since that would destroy Life's memory process. Digital codes can only be altered indirectly, i.e., through the rewriting of the messages for each new generation.

This corresponds quite closely to the role of the digital code in human life. The cookbook is obviously not going to be revised automatically because of a bad experience. An unsuccessful batch of meatballs cannot bring about changes in the recipe. And as a rule changes are not made to bad books or erroneous books. Instead these are simply not reprinted, but are supplanted by new books. Just like the genetic material, the books have to be rewritten. Neither the genetic nor the cultural texts are directly influenced by any prospective negation of events.

Another—though related—forte of digital codes such as DNA sequences and texts is that they are *not* contingent upon similarity with the "reality" in question. By virtue of this "alienness" or split (cf. chapter 1), the digital code frees human beings from the chains of actuality; that is to say it opens the door to creativity and imagination.

Within the pages of a book all the bonds imposed on us by the laws of Nature can be severed. Every imaginable kind of world can be unveiled, to cast a new light on the known world. Similarly, the organism's DNA-coded self-description paves the way for the combinatorics that are an integral part of sexual reproduction. Combining half a hen with half a cockerel will not produce a chicken. The messages in analogic code form are quite inflexible. Combining the two means resorting to the digital code. And, as everyone knows, combining hen and cockerel in chromosome form is as easy as pie.

It is this creative interaction between the two versions of the message that is entailed in code duality. Where the digital code takes care of the objective, conservative (or inherited) aspects of life, the analogic code—i.e., the actual organism—is designed to deal with the here and now; it represents the subjective, active, ecological protagonist in life.

It is, as it happens, a well-known fact that digital codes are not action-oriented. Words have no direct effect on the world. The reader's hair will not, I am happy to say, turn red simply because I write: "The reader is turning into a redhead!" And the rock is not going to roll away from the mouth of the cave just because someone says: "Open Sesame." Such faith in the power of runes or magic spells is usually described as superstition, and superstition is not something in which science is in the habit of indulging. So why believe that DNA plays an active role in the life process? One of the few to have grasped this point is the Harvard geneticist Richard D. Lewontin, who writes: "First, DNA is not self-producing, second, it makes nothing and third, organisms are not determined by it."[56] Why this memory lapse concerning the fact that "someone" has to interpret the digital code before it can be activated? Is this due merely to a horror of subjects, of "someone"? A horror of any non-quantifiable element in the natural world?

The first lines of Søren Kierkegaard's "Sickness Unto Death" run as follows:

> Man is spirit. But what is spirit? Spirit is the Self. But what is the self? The self is a relation that relates to itself, or else it is within the relationship that the relation relates to itself; it is not the Self that is the relationship, but that the relation relates to itself.[57]

Thus, according to Kierkegaard, a human being becomes a "self" insofar as it can, in a given action or choice context, stand back from the situation and evaluate its own relationship to it. Thus, a "self" presupposes a three-factor relation in which the individual refers both to the situation in which he finds himself and to his own presence in that situation.

The French philosopher Maurice Merleau-Ponty expounded a similar view. He rejected the simplistic idea of the subject as an ego which is the sum of its own experiences. Instead he stressed the importance of "non-coincidence." Monika Langer sums up Merleau-Ponty's standpoint thus: "Our reflections on time are themselves taking place within time, our reflections on subjectivity are themselves a part of our subjectivity. We can never become one (coincide) with ourselves precisely *because* of this non-coincidence gap."[58]

We are getting into deep water here. But I hope the reader has maintained enough sangfroid to note the strong connection between these two views of the self and the picture of the essence of life which I have presented in this chapter.

In the light of this I dare to conclude that it is permissible to use words such as "self" or "subject" of living creatures: Living creatures are self-referential, they have a history, they react selectively to their surroundings, and they participate in the evolutionary incorporation of the present in the future.[59]

With the emergence of code duality the world began to invent. In this chapter we have tried to discover how this came about. In the next chapter we shall see where it led.

5 OPENING UP

On the sensory universe of creatures:
The liberation of the semiosphere

Most of us are familiar with that faint pang, the slight shudder, that sometimes runs through us when we hear of someone being hurt in an accident. We do not even have to know the person concerned, and the little twinge we feel obviously bears no resemblance to the pain the injured person actually suffers. Yet it is as though our bodies have momentarily reached out to that other body.

Nevertheless, we would probably have to admit that other people are for the most part unfathomable. Even our own children remain something of a mystery. And on a visit to a foreign country, incomprehensibility leaps out at us from the merest detail. We have a hard enough job trying to see the world through other eyes, much less understand the world those other eyes see.

But is there really anything so strange in that? Would it not be fairer to say how amazing it is that we do have occasional flashes of insight, when we are able to see what others see or understand how they feel? While chimpanzees can obviously feel grief—can even die of grief, as in the case of Flint the young chimp who lost his mother—there is no record of a chimpanzee ever having tried to comfort a friend. Chimpanzees do not appear to have the capacity for perceiving that others might be in the same mental state as themselves.[60] The true wonder lies, like as not, in the fact that we humans have the ability to empathize at all.

It is one thing, however, to be able, now and again, to empathize with other human beings. Try to do the same with animals and you will soon find you are skating on very thin ice. "If a lion could talk, we could not understand him," wrote Wittgenstein.[61] Because we could never recognize the world in which it lives. Our language and our worlds are inextricably intertwined, living as they do in symbiosis with one another.[62]

And yet, on hearing birdsong on a summer morning it is hard not to feel that the birds must take something of the same kind of pleasure in it as we ourselves do. Having made a study of birdsong from a primarily aesthetic point of view, the biologist Joan Hall-Cragg also concluded that music is and always will be the exclusive privilege of birds and humans.[63] In a commentary on Hall-Cragg's work the philosopher Charles Hartshorne wrote: "Considering the enormous gap between the anatomies and lives of man and bird, it remains astonishing how much musical intelligibility the utterances of the latter have for the former."[64] But Hall-Cragg herself did suggest that the similarity between these two types of semiosis is due to the similarity of the demands which have, in the course of evolution, shaped them: The need to send a signal to a listener some way off. Well, yes . . . but I wonder whether that really covers it.

When one is as sound-oriented as all humans are, it might seem rather strange that, as Julian Huxley puts it, "the great majority of all animals are both deaf and dumb."[65] Apparently, most animals spend their days in a world that is, quite literally, mute. And even though crickets, cicadas, and other insects may "chirp" or "hum," they are in fact tone-deaf; that is to say they cannot differentiate between high and low notes. Which is why their "singing" possesses very little of what it takes to touch the human heart. A cricket can, by varying the volume and speed of its "chirping," produce as many as five different songs, each carrying its own message.[66] Not only that but these tiny creatures are totally dependent on these songs, if they are ever to recognize and mate with members of the opposite sex within their own species.[67]

The moth's sonic universe is even more limited. As far as can be gathered, it can pick up only one particular note, that being the note emitted at a frequency of twenty thousand hertz by a bat taking soundings to locate its prey—including, not least, moths. This note enables the moth to determine how far off its enemy may be and in what direction. If the bat is some distance away the moth will head directly away from the sound. If on the other hand the bat is close by, it will execute a number of abrupt and unpredictable movements.[68]

One of the pioneers of research into the "sensory worlds" of animals was the Estonian-born, German biologist Jakob von Uexküll (1864–1944). The essence of Uexküll's life's work is contained within the term *umwelt*, a term which I have already employed in previous chapters. Strangely enough this word can be traced back to a Danish poet, Jens Baggesen, who first used it around the year 1800.[69]

The way Uexküll saw it, animals spend their lives locked up, so to speak, inside their own subjective worlds, each in its own *umwelt*. Modern biology employs the objective term "ecological niche," that is to say the set of conditions—in the form of living space, food, temperature, etc.—under which a given species lives. One might say that the *umwelt* is *the ecological niche as the animal itself apprehends it*.

In *Bedeutungslehre* (*The Theory of Meaning*), published in 1940, Uexküll writes: "Standing before a meadow covered with flowers, full of buzzing bees, fluttering butterflies, darting dragonflies, grasshoppers jumping over blades of grass, mice scurrying and snails crawling about, we would instinctively tend to ask ourselves the question: Does the meadow present the same prospect to the eyes of all those different creatures as it does to ours?"[70]

The answer to this question is "No," and in this a modern biologist would agree totally with Uexküll. But while this "No" is quite central to Uexküll's whole view of nature, for biologists in general it seems almost to be a rather incidental oddity.

Uexküll uses a wild flower to illustrate his *umwelt* theory. How, he asks, does this flower fit into the *umwelt* of different creatures: (1) A little girl picks the flower and turns it into a decorative object in her *umwelt*; (2) an ant climbs up its stalk to reach the petals and turns the flower into a ladder in its *umwelt*; (3) the larva of the spittlebug bores its way into the stalk to obtain the material for building its "frothy home," thus turning the flower into building material in its *umwelt*; (4) the cow simply chews up the flower and turns it into fodder in its *umwelt*. Each of these acts, he says, "imprints its meaning on the meaningless object, thereby turning it into a conveyor of meaning in each respective *umwelt*."[71]

As far as Uexküll is concerned, a creature's specific *umwelt*, its subjective world, lies at the very heart of his analysis. "As the two parts of a duet must be composed in harmony—note for note, point for point—so," he writes, "in nature the meaning-factors are related contrapuntally to the meaning-utilizers in its life."[72] It is this contrapuntal interaction that moves him to call the flower bee-like and the bee flower-like, the spider fly-like, or the tick mammal-like.

This constitutes a polemic assault on mechanistic biology's reduction of Mother Nature's great masterpiece to an accumulation of inconsequential particles in a state of insignificant flux. The view whereby "the meadow consists of a confusion of light waves and vibrations in the air, finely dispersed clouds of chemical substances and chain reactions which control the various objects on the meadow."[73]

Uexküll counters this by stating that "colors are light waves which have become sensations: This means that they are not electrical stimuli, acting on the cells of the cerebral cortex, but the *ego qualities* of those cells."[74] Apart from that one term—*ego quality*—this is still more or less in line with modern biological thinking. There is no such thing in this world as an objective perception of the property "redness." It is a sensation constructed by the organism's brain cells. The human eye will, for example, perceive a white dot as being red, if it is presented on a green ground (the eye having produced the complementary color to green, i.e., red, in the white area).

Uexküll has a fondness for employing musical metaphors to illustrate his point of view. Thus the expression "ego-quality" is a modernized version of his own "ich-ton," that is to say "I-tone," which he defines in the following fashion:

> Imagine that we had a number of living bells, each capable of producing a different tone. With these we could make chimes. The bells could be operated either mechanically, electrically or chemically, since each living bell would respond to each kind of stimulus with its own "ego-quality" [ich-ton]. . . . Chimes composed of living bells must possess the capacity to let their tune resound, not only because they are driven by mechanical impulses, but also because they are governed by a melody. In this manner, each ego-quality would induce the next one, in accordance with the prescribed tone-sequence created by the melody.[75]

Elsewhere, he compares the genes in a fertilized egg to the keys on a piano, upon which formative melodies may be played.[76]

Uexküll committed the crime of not believing in evolution. He found it absurd to assert, as those Darwinists among his contemporaries did, that ". . . this romantic duet that sounds throughout the whole of the living world in thousands of variations can have come into being totally without design."[77] Thus he took up a stance which, as the twentieth century progressed, became more and more untenable. And this is probably one of the chief reasons that his work has been pretty much ignored in biological circles. This, despite the fact that he had a profound influence on Konrad Lorenz's thinking and was actually, indirectly, the founder of ethology, the study of animal behavior.

But it is not a matter of harking back to the past just to find heroes to worship. History shows that everyone makes mistakes of one kind or another. It also shows, though, that important discoveries are forever being obscured by new trends in thinking and, thus, consigned to oblivion. That is what has nearly happened to the *umwelt* theory.

Jacob von Uexküll's line of thought is, at heart, semiotic, or biosemiotic,[78] though he himself never used these terms, just as it is highly unlikely that he was familiar with Peirce's work. His whole point was that neither the individual cells nor the organisms are passive pawns in the hands of external forces. They create their own *umwelt* and in so doing become a subjective part of Nature's grand design.

And this is where Uexküll's theory becomes quite definitely a thorn in the flesh of modern biological thinking. According to the view currently in vogue, evolution occurs through an external weeding-out process, or natural selection. As opposed to the organism, selection is a purely external force while mutation is an internal force, engendering variation. And yet mutations are considered to be random phenomena and hence independent of both the organism and its functions. By this token the organism becomes, as Claus Emmeche says, "the passive meeting place of forces that are alien to itself. The irony is that this renders it irrelevant to evolutionary biology."[79]

This image of the organism as a passive object for the process of change appears to be yet another example of this urge to hack away at Nature, trimming its edges to render it quantifiable. On the face of it this seems odd—running counter to all our instincts for what goes on in the natural world. But it also leaves itself open to academic biological criticism, as two heavyweights within the field of evolutionary biology, Richard Levins and Richard Lewontin, have recently pointed out.[80]

The crux of their argument is that organisms do themselves interpret and respond to their environment. As a response to external change an animal might, for example, home in on selected areas of its habitat. One dramatic example of this was shown in the case of the locust, which altered not only its behavior but also its anatomy, as a response to climatic disorder. Organisms can also alter the environment in which they live. Plants, for instance, alter the physical and chemical makeup of the Earth as they grow. Although the organism is of course shaped by the interplay between genes and environment, the environment is also to some extent shaped by the organism. The organism plays an active part in its own construction.[81] Critics of neo-Darwinism freely admit that natural selection at the individual level is a central factor in evolution. As an explanation, natural selection does, however, leave something to be desired, since it too requires an explanation within a wider thermodynamic or cosmological perspective.[82] This selection is a process which has itself been selected, so to speak,[83] and that principle of selection—which, in fact, comes into play at a more fundamental level than the process referred to by biologists as "natural selection"—needs to be brought to the fore in evolutionary theory. But this raises a problem implicit in the idea of selection, one which biologists only succeed in hushing up because, as a rule, they refuse to treat the question of selection theoretically, i.e., as divorced from the *ad hoc* circumstances under which, in concrete terms, selection always occurs. The problem is that any process of selection presupposes an intention or a ground rule that determines what will be selected. This is exactly what Peirce called "the tendency to take habits" (see chapter 2).

It is quite feasible that the reluctance to accept such a conclusion—which does, after all, imply some sort of "directedness" or inherent design in our universe—is precisely what causes most biologists to dismiss any criticism of neo-Darwinism. But the old head-in-the-sand ploy, besides being not exactly creditable, does not look like such a safe bet in the long run. At some point the neo-Darwinists are going to have to lift their heads and face the fact that the problem of subjectivity cannot simply be spirited away. This world is full of subjects and something must have created them. But latent within that "something" there must, inevitably, be "someone." Subjectivity has its roots in the cosmos and, at the end of the day, the repression of this aspect of our world is not a viable proposition. We would do better to study it and try to reach an understanding of it.

We need a theory on organisms as subjects to set alongside the principle of natural selection, and Jakob von Uexküll's *umwelt* theory is just such a theory. Ironically, however, it is only through integration with the theory of evolution that the *umwelt* theory can truly bear fruit.

Such irony is, however, not uncommon in the history of science, where the controversies of one period are often seen to be quashed at a later date through synthesis at a higher level. One renowned example concerns the dispute, at the end of last century, over fermentation. The chemists (with Justus Liebig at their head) believed this to be a chemical process; the microbiologists (with Louis Pasteur as their spokesman) claimed that it was a microbial process. In retrospect we can now see that both parties were right because, although fermentation is indeed usually carried out by yeast cells, the yeast cells derive their fermentative ability from enzymes which can also be made to function outside the cell.

As soon as we start to interpret the *umwelt* theory in historical terms the pieces begin to fall into place. Because the *umwelt* theory tells us that it is not only genes, individuals, and species that survive, but also—and perhaps rather—patterns of interpretation. A creature's *umwelt* can be seen as the conquest of vital aspects of events and phenomena in the world around it, inasmuch as these aspects are continually being turned—by way of the senses—into an integral part of the creature. The *umwelt* is the representation of the surrounding world within the creature.

The *umwelt* could also be said to be the creature's way of opening up to the world around it, in that it allows selected aspects of that world to slip through in the form of signs. Even the moth's *umwelt*, otherwise so silent, has kept just one chink open to admit the bat's few fatal soundings. The specific character of its *umwelt* allows the creature to become a part of the semiotic network found in that particular ecosystem. It becomes part of a worldwide *horizontal semiosis* (cf. chapter 3).

The advantages of being in possession of a sophisticated *umwelt* are many and various. The most important of these is perhaps *anticipation*, the possibility that the *umwelt* offers the organism of predicting events which it can then defend itself against or make use of in some other way. Horizontal communication is, moreover, one of the prerequisites for the social life of a population. It allows for greater, and perhaps more especially, more dynamic social complexity. Highly developed horizontal semiotic activity is of course also essential if an animal is to become involved in learning processes.

As evolution progresses, the horizontal semiosis becomes detached from the vertical semiosis through the development of creatures with increasingly complex *umwelt*s. This "trend" can be seen as a gradual delegation of the authority to decide how life is to be tackled from the genetic material to the organism itself. The anticipation inherent in the genetic material becomes more and more the anticipation of the anticipatory skills which organisms must possess if they are ever to reproduce successfully. What emerges is a sort of Chinese puzzle— representations within representations. As each individual's ability to interpret signs in its surroundings becomes more and more far-reaching, the mating game becomes woven into complex semiotic patterns that are clearly linked—though only to a limited extent—with the domain of strict genetic directions.

This semiotic network, Uexküll's "contrapuntal duets" in their entirety, constitutes an emergent level, in the sense in which this term was employed at the end of chapter 3. Let us call it the *semiosphere*. The semiosphere imposes limitations on the *umwelt* of its resident population in the sense that, to hold its own in the semiosphere, a population must occupy a "semiotic niche." To put it another way, it has to master a set of signs, of a visual, acoustic, olfactory, tactile, and chemical nature, by means of which it can control its survival in the semiosphere. An ecosystem is usually defined by the currents of energy and matter that bind individual populations within a network together. But one could ask what role the semiosphere's own network of communicative relations does in fact play? It seems very likely that the semiotic demands made on an organism are vital to its success. In which case one can never hope to understand the dynamic of the ecosystem without allowing for some form of *umwelt* theory.

Since Darwin's day biologists have been debating whether evolution can reasonably be said to have had a particular direction. To begin with there was talk of progression, since humanity—which of course constituted the ultimate stage—was quite naturally considered to be superior to other animals. But the progression theory was generally viewed as being unscientific, entailing as it does a value judgment as to what is best. One objection to the idea of evolutionary progression might be that bacteria, their simplistic design notwithstanding, have undoubtedly come off well in evolution. And what scientific grounds could there possibly be for claiming that something complex is a step up from (i.e., better than) something simple?

A typical body of opinion in modern biological circles would have it that every species in existence today is, by definition, best because it has managed to survive—there are no other criteria.

And yet it is not "unlawful" to debate whether evolution can be said to have a *direction,* whether there is one aspect of life that has been systematically reinforced throughout the history of the natural world. Various candidates for such a role have been proposed—for example, the wealth of species living on the Earth, the volume of biomass relative to one unit of photosynthetic energy, or simply longevity. But the most popular notion would seem to be that evolution has generally tended to produce ever more complex life-forms. Now here one has to be careful not to fall into the trap of saying that complexity is the purpose of evolution. One can, however, assert that natural processes happen to be arranged in such a way that they will ultimately result in the creation of greater and greater complexity.

Such a view also seems to be borne out by the physical chemistry.[84] Furthermore, this seems instinctively to tally with the (somewhat questionable) notion, popular among environmentalists, that a complex ecosystem is better or more stable than a simple ecosystem. This appears to be founded on the notion that we ought to follow the same course as evolution—a thesis which undeniably loses a lot of its charm once we realize that humanity is hardly likely to be the ultimate goal of evolution.

The problem is, however, that it is not exactly clear just what this complexity amounts to. According to the evolutionary biologist Daniel McShea, who has written an article summarizing this subject, it is now more or less agreed that the morphological complexity of a system is determined by the number of different parts of which it is comprised and the greater or lesser irregularity of their arrangement. A complex system is therefore heterogeneous, detailed, and lacking in any particular patterns.

McShea did, however, have to round off his review of literature on the subject by saying that, despite what common knowledge would have us believe, there is hardly any empirical evidence to support the theory that complexity, in the above-mentioned sense, has grown greater in the course of evolution.[85] Apropos of this, he quotes the distinguished palaeontologist George Simpson as saying as far back as 1949: "It would be a brave anatomist who would attempt to prove that Recent man is more complicated than a Devonian ostracoderm." (Ostracoderm: a species of fish, to which the trunkfish belongs, that was in existence between three and four hundred million years ago.)

From a biosemiotic point of view it seems hardly surprising that a concept of complexity as dull as McShea's should come to grief. Over and above the morphological complexity one ought obviously to allow, at the very least, for behavioral and social complexity. But even at this point one might well run into problems with converting this complexity into numerical form, and so might feel that it would be better to ignore behavioral patterns and simply maintain that the level of complexity does not increase.

But my point is of course not so much that the spotlight should be turned on morphology and behavior, as that attention ought to be directed at what is, in fact, the essence of the entire life process—life as semiosis. The most pronounced feature of organic evolution is not the creation of a multiplicity of amazing morphological structures, but the general expansion of "semiotic freedom,"[86] that is to say the increase in richness or "depth" of meaning that can be communicated: From pheromones to birdsong and from antibodies to Japanese ceremonies of welcome.

I should be very surprised if the driving force behind evolution did not prove, at the end of the day, to be the self-same creativity and flexibility that are accorded to those systems engaging in ever subtler forms of semiotic interplay. The anatomical aspect of evolution may have controlled the earlier phases of life on Earth but my guess is that little by little, as semiotic freedom grew, the purely anatomical side

61

of development was circumscribed by semiotic development and was thus forced to obey the boundary conditions placed on it by the semiosphere.[87]

I had to think long and hard before choosing to speak of semiotic "freedom" rather than semiotic "depth." It was not an easy decision to make, since freedom is—as we have already seen in chapter 3—a rather ambiguous term. Semiotic freedom refers not only to the quantitative mass of semiotic processes involved but even more so to the quality of these processes. We could perhaps define it as the "depth of meaning" that an individual or a species is capable of communicating.

Over recent years it has become quite clear that some kind of term along the lines of "depth" is required in the communication sciences, to supplement the term "information." This whole dilemma is quite brilliantly dealt with by Tor Nørretranders in his book (*Mærk Verden*).[88] But I am afraid that I do not agree with a couple of the most crucial points made in what is, otherwise, a splendid book. First of all, I do not believe that the transfer of the information concept to biological and psychological systems can be achieved as painlessly as his book suggests.

Even though this takes us off at a tangent, there is a very basic point here which we must clear up before I dare proceed with this book's very own fragile and irreversible semiosis. The sad thing is that the term "information," which has served physicists so well, has now broken free of physics and become mixed up, in the most bewildering fashion, with the everyday version of the word information. For the rest of this book I will therefore use the word "information" in quotation marks to denote the physical "information" while the same word without quotation marks will mean just what information has always meant.

And what might that be? This point can be clarified with the help of a little etymology. The word information comes from the Latin *informare*, meaning to give material form. Oddly enough the word "form" was the Romans' mangled version of the Greek word "morf." And morphology, of course, is the study of form—the Romans just happened to transpose the m and the f. So in days of old being informed meant being "formed" or "cultivated." And what was to be formed was our intellectual awareness. Information was something that got us into shape mentally.

Thanks, however, to the twentieth century's urge to reduce everything to atoms, information gradually came to mean almost the same as isolated facts, chunks of knowledge. And that is more or less what it meant when the physicists commandeered it.

The unfortunate thing about the physicists' concept of "information" is that it no longer refers to a person or to any other subject. The "information" of the physics world is not something that "someone" has; it is there, in the world, quite regardless of whether "someone" is also there. Defined thus, in purely physical terms, this term cannot encapsulate what is—for human beings—the all-important quality inherent in the fact that information can be relevant or irrelevant. Physical "information" takes in the whole kit and caboodle.

I will not be using this term in the rest of this book so there is no need to go into it in greater depth here. The point to note is the consequence which it has, namely that from a mere mortal's point of view "information" almost seems to be some sort of intellectual junk that has to be got rid off. In a supermarket, for instance, we are usually quite happy to be presented with a total for what we have to pay. But then the prices of all the individual purchases lose their "information." So normally we are not interested in "information" but only in "information" that has been sifted or pre-processed. For the same reason, a monkey writing 224 pages on my word processor would produce a piece of text containing considerably more "information" than this book. Because the monkey would not differentiate between the signs. All modesty apart, I would have to say that I consider such use of language quite absurd.

Strangely enough this absurdity does not seem to worry the physicists, who unceremoniously persist in lecturing the rest of us on the "true" definition of information. But it is hard to understand why the rest of the world should defer to the physicists' terminology rather than the other way around. The simplest solution would of course be for the physicists to find another word to cover that phenomenon which has at some point mistakenly been dubbed "information." The explanation may well be, however, that mathematics and physics are Greek to most people, and we all humbly doff our caps to Greek—or Latin for that matter.

Now, however, rather than correct their mistake, the information theorists are endeavoring to retrieve what was lost with the aid of new terminology. And since the problem with the "information" of the physics world lies in sifting through it to get at the miniscule amount required only to throw the rest away, then perhaps they should try to come up with some kind of yardstick for distinguishing between "junk information" and processed "information." Such a yardstick would compensate for the disparity between "information" and information.

This is where the term "depth" comes in. Information can be of greater or lesser depth. In 1985 Charles Bennett of IBM introduced the term "logical depth." "Logical depth can loosely be defined as the number of steps in an inferential process or chain of cause and effect linking something with its probable source," writes Bennett.[89] The logical depth of a statement is, therefore, the expression of its meaning, its worth. The more difficult it is for the sender to arrive at a statement the greater its logical depth. The more "working-out time" he has used—in his head or on a computer—the greater its value becomes because he has saved the recipient from having to do this work.[90]

But here there would appear to be a problem, because the very mention of "meaning" and "worth" obviously brings us back to the human—or, generally speaking, the subjective—world. The question is how, in this case, did we get from the world of physics to the world of man? And the answer is, we did not. Logical depth still relates to a computer, in terms of the number of calculations that are required. "Worth" is measured in computer time which, I must emphasize, is a ridiculous gauge of the worth, say, of a collection of poetry.

In a visionary article published in 1988, physicists Seth Lloyd and Heinz Pagels try to get around this objection. They suggest that, rather than measure the logical structure of an object, one ought to work out a method of measuring "the most likely way for an object to have come about."[91]

I would agree that—in theory—this sounds like a promising project. But, as far as the emergence of human beings or the shaping of animal *umwelts* are concerned, the evolutionary history poses the one problem that anthropologists, psychologists, and biologists have developed whole screeds of terminology to tackle. Without, it must be admitted, much luck. And yet I find it hard to see how some smart physical rule of thumb for discarded information could run rings around years and years of scientific achievement in those fields.

No matter how intuitively convincing Lloyd and Pagel's idea may be, the main difficulty lies in translating this into something quantifiable. This feat has not yet proved possible, and why on earth should it?

A choice still exists between creating certain unrealistic and oversimplified models of reality—which do, however, have the advantage that they can be quantified—or creating realistic models which, as such, cannot be quantified. Lloyd and Pagels have opted for the latter and all credit to them for that.

In the foreword to his book *Acts of Meaning* the psychologist Jerome Bruner writes: "If we take the object of psychology (as of any intellectual enterprise) to be the achievement of understanding, why is it necessary under all conditions for us to understand *in advance* of the phenomena to be observed—which is all that prediction is? Are not plausible interpretations preferable to causal explanations, particularly when the achievement of a causal explanation forces us to artificialize what we are studying to a point almost beyond recognition as representative of human life?"[92]

I find both Bennett's and Lloyd and Pagel's ideas thought-provoking. They can be of help in defining the way in which we perceive complex systems. The problem does not arise until the moment one attempts to artificially formalize concepts by turning them into quantitative "yardsticks." While this sort of approach may work in the sphere of physics and technology, there is no guarantee that it will work in the sphere of life and the human psyche. And there is more than a hint of arrogance in believing that the work of generation after generation of learned individuals can be dismissed out of hand.

Always the danger is, as Bruner points out, that a quantitative measurement falsifies or distorts whatever it is measuring. How often have we not seen how this results in whatever was originally to be measured being eclipsed by the measurement itself? As in the case of the intelligence quotient—originally designed to ascertain which children were in need of extra teaching but soon used to prevent "inferior" persons from entering the USA, and to justify the sterilization of women who had been declared mentally retarded.[93]

The strange thing about this entire saga of the information concept is that a definition of information was provided long ago by Gregory Bateson, one which describes this word's colloquial meaning to a T: "a 'bit' of information is definable as a difference that makes a difference," he writes.[94] The essence of this definition is that information is something which is generated by a subject. Information is always information for "someone"; it is not something that is just hanging around "out there" in the world. For instance: If I happen, one evening, to hear a blackbird burst into song, I might look up into the tree to try and catch sight of it. In other words, the variations in sound reaching my ears prompt my brain to produce a piece of information to the effect that there must be a blackbird somewhere close at hand. For the moth clinging to a nearby wall, on the other hand, no information whatsoever is generated. The blackbird's song is a difference that makes absolutely no difference to it. Ergo, no information. And my small son might well contrive to say "bird," but not "blackbird."[95] He has, in other words, produced another piece of information from the same sound.

The annoying thing about Bateson's definition is that it cannot be used to quantify information. Information is associated with an intentional creature of some kind or another, whether it be an amoeba registering a difference in nourishment levels and reacting by extending a pseudopodium toward the spot where the pickings are richest, or a human being seeing a ripe fruit on a tree and stretching out a hand to pluck it. Or—to put it another way—information is based on interpretation and, in this sense, corresponds to signs as defined by Peirce.[96]

I opted for the expression semiotic "freedom" rather than semiotic "depth" in order to save giving the (false) impression that we are dealing with a quantitative term on a par with logical depth. My aim, in calling it semiotic freedom, is to establish the creative dimension of the semiosphere. But just because semiotic freedom cannot be weighed up in the same way as a quantitative measurement does not mean to say that this term could not be defined—according, for example, to principles similar to those involved in the term *logical depth*. For the moment this remains open to question.

In this chapter I have dealt with horizontal semiosis, as it occurs at the ecological level. From the individual's point of view this involves something which could be called *exosemiosis*,[97] i.e., sign processes that occur between (and outside of) organisms. But as we have seen, these processes were dependent on animals developing, in the course of evolution, the ability to be receptive to the signs on offer from their surroundings. The purpose of this receptivity was to arrive at internal representations of whatever lay outside. But the fact of the signs' penetrating the body inevitably leads to the execution, internally, of a wealth of sign processes—an endosemeiosis.[98] Of course the skin may not seem to be an obvious barrier to semiotic processes. So we will have to complete the picture by taking a closer look at the relationship between external and internal signs.

6 | DEFINING

The mobile brain: The language of cells

Men who suspect their partners of being unfaithful have been found to produce more sperm cells and have more prolific ejaculations than men who harbor no such suspicions. Experiments were carried out by doctors in Manchester in which test subjects were split up into two groups according to their confidence, or lack of same, in their partner's fidelity. The number of sperm cells produced by their ejaculations were then counted and compared. Dorion Sagan and Lynn Margulis, who discuss such experiments in their book *Biological Striptease,* believe that "this unconscious control of sperm production is clear proof of the existence of sperm rivalry in human beings."[99] This sperm rivalry presupposes that females will mate with several males during the one cycle, in which case millions of sperm from two or more ejaculations will be swimming against one another in a race to reach their goal. Sperm rivalry implies therefore that the competition among males to beget offspring is controlled by the ability of their sperm to beat the sperm of other males.

If this is correct, then it makes good biological sense to produce more sperm as the competition gets fiercer. But to say that it is expedient is not an adequate explanation of this phenomenon. We also need to ask how such a thing can happen. In other words: How do the man's testicles know that his wife is fooling around, or rather, that the man thinks she is?

This is an awkward question. After all, how can one answer it without admitting that purely mental phenomena such as fancies or dreams can have an effect on physical phenomena? And having admitted that, how can one then carry out meticulous, double-checked medical experiments that do not place any credence in fancies and dreams? Or, to put another way, what is left of that much-vaunted objectivity?

Questions of a similar nature seem to be cropping up more and more often. And any sensible doctor now knows that a person's psychological and social circumstances play a crucial part in determining his or her general health prospects and susceptibility to disease. The only problem is that the theoretical structure of medicine is founded on a biological model of human beings which systematically prohibits these psychological and social circumstances from being brought into play. Because biology, like all other sciences, is based on Descartes' old dualism, which does not allow any inner link between the spiritual and the physical sides. So when medicine and biology decide to take psychological phenomena seriously, this can only be done by reducing said psychological phenomena to mechanical occurrences.

To take a typical example: There is some talk of alcoholic or schizophrenic genes. While not denying that inherited factors might quite feasibly have a part to play in personality disorders, it must be stressed that this viewpoint turns these psychological phenomena into rigid caricatures. Experience tells us that the great writers were barely able to plumb the rich depths of the psychic domain, so why should we imagine that alcoholism can be summed up by an explanation as simple as "an alcoholic-gene"? No wonder that medicine has been unable to make any serious contribution to the understanding of the psychosomatic aspect of illness. As a matter of fact it would be truer to say that it has prevented us from reaching an understanding of this problematic area.

This is an intolerable state of affairs (at any rate for everyone apart from those whose places in the medical world's pecking order depend on it). And the greatest benefit of the biosemiotic approach may be that it can get us out of this mess. Because inside the body, too, processes are in fact occurring which can best be understood semiotically, processes which make it possible to understand how the body can become "minded" and how the mind can become physical.

The key to understanding the semiotics of the body—the endosemiotic level—lies in the term *receptor*. This term was originally used to denote the nerve endings found within the sensory organs. But today it is also applied, as we shall see a little later, at the molecular level. In both cases it is used to describe "instruments" for picking up and responding to signals from outside, be they signals from the organism's surroundings or the surroundings of an individual cell. And in both cases the signal is seen to cross a barrier, be it the skin or the cell membrane. Finally, in both cases the signal is translated into a form that makes sense to the system on the inside of the barrier—the animal or the cell.

A look at the sense of balance in vertebrates provides us with an example of how the sensory receptor works. In this case the receptors are comprised of tiny "hair cells," i.e., cells with a little hair-like tuft (a cilium) at one end and their other end linked to a nerve. These hair cells are arranged in rows on the inside of the utriculus, a chamber of the inner ear, which is full of a gelatinous substance. And floating around in this substance are a number of pebble-like concretions of calcium carbonate. When these crystals collide with a hair, causing it to bend, the hair cell transmits a nerve impulse. If the animal is at rest, the pattern of the bent hairs will reflect the head's position in the field of gravity. If it is in motion the "sloshing" of the crystals will reflect every change in its pace.

In this instance, position and speed are first translated into pressure (on the hair cells) and then into electrical impulses, nerve impulses, which are transmitted along the nerve fibre until they reach the brain. Inside the brain these nerve impulses are checked against hosts of other

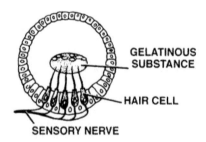

FIGURE 10. *Sensory nerve, hair cell and gelatinous substance. The hair cells in the inner ear. Tiny crystals float around in the gelatinous substance. When these hit the hairs, making them bend, nerve impulses are transmitted to the brain. From the patterns formed by the bent hairs the brain can work out the position of the head in the gravitational field.*

nerve impulses coming, for example, from some of the other organs governing the balance, or from sight or hearing, and in this way each impulse is absorbed into the animal's actual actions.

So the receptors are directed at the surrounding world, which they allow to slip through to us. Which fragments of the surrounding world manage to slip through is something that is determined by the disposition of the receptors—another species-specific characteristic. And what gets through to the brain is still not the surrounding world as such—clearly, the telephone I am looking at right now doesn't have a double inside my head! Again, all we have to work with is a set of signs, yet another coded version—with the code here being the pattern of electrical impulses. It looks as though deciphering this code will remain, for some decades to come, an impossibility for any entity but the brain itself, that brain which is inundated around the clock with signs both from the world outside and from the inner world—nerve impulses, some of which it selects and some of which it rejects.

What we can say, however, is that the world we see, feel, hear, etc., that is to say the world of which we are aware, is a heavily processed world. Our sight is a good illustration of this. The current view on this point would appear to be that, through the lense of the eye, an animal or a person fixes an image on the retina. This is then transformed by the nervous system, much in the same way as the voice in a telephone conversation is transformed into electrical impulses. This input enables the brain to reconstruct the three-dimensional representation of the real world which we call sight.

But this is far too simplistic a picture. On its way from the retina to the vision center in the cerebral cortex every single nerve impulse passes through a tiny area on the underside of the brain known as the CGL (corpus geniculatum laterale). In the CGL, nerve cells from the retina come into contact with nerve cells from many other parts of the brain—including the vision center in the cerebral cortex. In fact for every nerve cell from the retina there are a hundred nerve cells from other areas. So what actually gets through to the cerebral cortex's vision center has to a very great extent been pre-processed. A process that is carried out quite unbeknown to us.

The projection from the retina cannot, however, according to biologists Humberto Maturana and Fransisco Varela, be compared to an incoming telephone line. They write: "It is more like one voice being added to many other voices in a lively family discussion where reaching agreement on, for example, action to be taken will not depend on what *one* particular member of the family has to say."[100] So it would not be correct to say that what is seen is dependent on the seeing eye, since it would be truer to say that it is the brain that does the seeing.

The attentive reader will have noticed that this brings us back to Jakob von Uexküll's "ich-ton" or "ego quality," albeit in an updated version. Our *umwelt* is a helpless victim of the reconstruction which the receptors and the brain have conspired to create. When struck, the individual nerve cell just emits its own note, while the brain makes use of the world in composing its own melody. Maturana and Varela introduced the term *autopoiesis* (i.e., self-generation) to describe this self-reliance, which they saw as being a general trait in living systems at all levels. A unique feature of living organisms, write Maturana and Varela, is that "the nature of their organization is such that their only product is themselves. The being and doing of an autopoietic entity are inextricably linked."[101] While not wanting to affiliate myself with the philosophy in which Maturana and Varela have wrapped their theory of autopoiesis, I certainly find the concept of autopoiesis extremely useful.

Just as the body has its sensory apparatus, so each and every cell has its "receptors," i.e., proteins, which have the job of collecting and translating signals from around the cell. These signals are usually chemical in nature, since each receptor recognizes or responds only to one or a very few molecules.

But we will have to make a slight detour to look at what might be called the question of scale. It is almost impossible to imagine what a hive of activity our bodies would be seen to contain if, rather than looking at it from our own level on the scale, where distances are measured in meters, we were to drop down to the cell's level. Here, distances are normally measured in μm, i.e., thousandths of a millimeter. Even though the body feels reasonably firm and well-defined, under the skin it is in fact all but fluid. And everything is in motion. The majority of molecules wear out and are replaced by others within a few months, and within seven years the very fabric of the bones will have been completely renewed.

In an earlier chapter I talked about Mixotricha, the termite's intestinal protist, with the five hundred thousand bacteria swarming across its surface. Now that does sound rather a lot, when one considers that Mixotricha is a single-celled eukaryotic organism. But in our bodies five hundred thousand cells take up no space at all. If, for example, they were liver cells they would just about cover a pinhead. Which only goes to show that an adult human being is made up of a pretty astronomical number of cells. I have no figures. But the nervous system alone is said to contain something in the region of 10^{11} cells, that is to say a hundred billion cells—twenty times as many cells as there are people on Earth.

If we drop down yet another notch on the scale, to the world of proteins, we have to measure in nanometers, with 1nm representing one millionth of a millimeter. A protein is a long, filiform molecule shaped like a tangled ball of wool. Despite its bewildering mass of loops and coils, the protein molecule actually folds up into an extremely well-defined three-dimensional structure, with a standard diameter of between five and 10 nanometers. It might be easier to picture the scale at which we are working here, if we think of the cell as a city like Manhattan. The nucleus of the cell would perhaps cover the area of Central Park while the membrane around the cell would be situated about as far from the center of the cell as the Bronx is from Fifth Avenue. In such a city, a protein molecule would be the size of a family car. And it would have almost less freedom of movement within the cell than a car has in Manhattan. Membranes and stop signs steer every single protein round and about in well-ordered patterns.

It is only when life-forms are seen in this perspective—of proteins, swarming around inside cells, swarming around inside bodies, swarming around in the hectic world outside—that one begins to understand the full extent of the coordinative puzzle that every living creature is charged with solving every single second of its life. That such coordination can be achieved is due to a regulating mechanism that is, to a great extent, decentralized—by virtue of communication systems between elements that are capable of interpreting their own situation, organs, tissue, cells, and even intracellular structure.

With this in mind let us now return to the receptors on the surface of the cell. By now, the reader will scarcely be surprised to hear that the cell's outer membrane is home to millions of such receptors. Generally speaking, all of these receptors can do two things. They can recognize and bind themselves to one specific molecule and they can alter their spatial structure. There is, for example, one kind of receptor which specializes in recognizing a particular signal molecule sent out by its neighboring cell. Another recognizes the protein that conveys iron around the body, a third may latch on to a particular hormone and so one. Recognition and bond are based on the "hand in glove" principle. The receptor's spatial surface is covered with hollows and projections which form a sort of cast of the corresponding spatial surfaces on the signal molecule. The bonding process also receives a helping hand from the electrical charge on the molecule surfaces. In short, the receptor is, both spatially and electrostatically, an analogic code of the signal molecule.

This bonding process causes an abrupt change in the spatial structure of the receptor. Usually, the receptor pokes its one end out through the cell membrane in order to make contact with the signal molecules, while the other end stays on the other side of the membrane. And it is here, on the end of the receptor molecule inside the cell, that the steric alteration takes place. This then triggers off an internal reaction of some kind within the cell.

As an illustration of this let us take a look at the receptor that recognizes the body's most important stress hormone, adrenalin (fig. 11). As a rule, adrenalin is secreted by the adrenal medulla during periods of exertion or as a consequence of emotional ups and downs. The adrenalin is carried with the blood around the body, having the general effect of increasing the blood pressure and releasing blood sugar from the body's store of glycogen (a high molecular carbohydrate that can be broken down into sugar). One of the areas where this occurs is in the liver cells, on whose surfaces receptors are all ready and waiting to seize the adrenalin molecule.

Bonding with the adrenalin molecule instigates the following complex sequence of events (related here solely as a means of illustrating how the process works): (1) On the inner side of the membrane a steric change in the receptor molecule occurs. This leads to (2), a so-called G protein, a neighbor of the receptor on the inner side of the membrane, altering its structure. This structural alteration entails (3), the G protein's relinquishing the grip which it has, until now, retained on the low molecular compound GDP and binding itself instead to a close relative of GDP, GTP. (4) The G protein is now split into its subelements, and (5), one of these (the alpha element) filters its way along

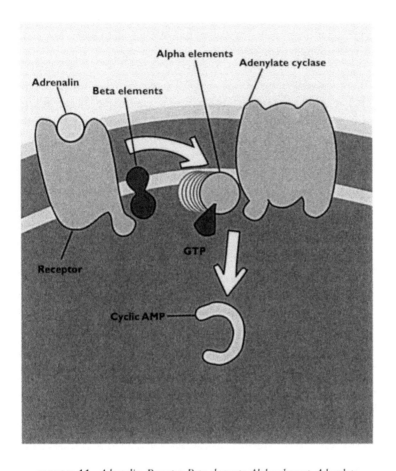

FIGURE 11. *Adrenalin, Receptor, Beta elements, Alpha element, Adenylate cyclase, GTP and Cyclic AMP. The adrenaline triggers the breakdown of glycogen into sugar. Alpha and beta elements combine to form the G protein. On receipt of an adrenalin molecule the G protein splits up into its sub-elements; the alpha element moves off to join the enzyme adenylate cyclase, which proceeds to produce cyclic AMP. Cyclic AMP then instigates the breakdown of glycogen into sugar. Described more fully in the body of the text.*

the membrane until it runs into an *effector* known as "adenylate cyclase" which (6) is thereby activated. (7) Adenylate cyclase prompts the production of a low molecular signal molecule called cyclic AMP. (8) Cyclic AMP now triggers the intricate chain reaction by means of which the high molecular carbohydrate glycogen is broken down into sugar. This sugar is then conveyed out of the cell, to make its way to hard-working muscle cells in dire need of energy.

The most vital factor in this complex sequence relates to the GDP-GTP team. The G protein can in fact only activate the effector as long as it is still "loaded" with GTP. But the GTP automatically breaks down ever so slowly into GDP and all of the alpha elements will gradually lose their GTP. When that happens they are reunited with the abandoned part of the G protein, the breakdown of glycogen comes to a halt and everything is back to square one—until a new adrenalin molecule comes along. It would seem that the receptor works through the G protein rather than stimulating the effector—the adenylate cyclase enzyme, that is—directly, in order to capitalize on this "delayed action" effect.

But it is quite conceivable that other benefits are also to be gained by taking a detour through the G protein. Because it just so happens that different receptors can have an effect on the same G protein, just as different G proteins can occasionally be utilized by the same receptor. This enables the cell to vary its response to a given signal. As Maurine Linder and Alfred Gillman wrote in *Scientific American:* "The ability of receptors, G proteins and effectors to interact with more than one species of molecule inside the cell also means that a cell can make different *choices* from time to time—now sending a signal down one pathway and now directing it along a somewhat different route" (my italics). Note how even in *Scientific American* leading scientists are now tempted to credit cells with the ability to make a choice, as though the cell were a *bona fide* subject.[102]

I must apologize for subjecting the reader to all this minutiae, the understanding of which demands a dreadfully persnickety level of concentration of the uninitiated. But the fact of the matter is that here we find ourselves with a phenomenon which, to my mind, occupies what is, theoretically speaking, a quite crucial position: right at the crossover point from biochemical processes to actual signs, where an interpretative body is involved.

At this purely biochemical level, literary scholar Frederik Stjernfelt has suggested the use of the term "categorical perception."[103] This term was originally introduced to describe a human being's ability to differentiate between different language sounds. This is actually a quite amazing skill, which human beings are born to master with no difficulty whatsoever. As far back as the 1950s it was demonstrated that we humans are very slow to identify non-linguistic sounds, recognizing—at best—seven to nine sounds per second. But when it comes to the spoken word, we are capable of differentiating correctly between sounds even when these speech sounds (which, by and large, correspond to individual letters) are usually being fired at us at a rate of fifteen to twenty-five per second.[104]

This tour de force is all the more impressive when one considers how little there is to choose between the sounds. If, for example, we begin a word with "b" or "p," in both cases we suddenly part our lips. This results in a *burst,* a sudden sharp sound. If the voice comes in less than twenty-five milliseconds (twenty-five thousandths of a second) after that smack the sound is heard as a "b." If, on the other hand, it comes in more than twenty-five milliseconds after the smack it is heard as a "p."[105] But not only do we differentiate at lightning speed between the most incredible details in the sound picture, we also compensate automatically for any possible difference in the length of the speaker's pharynx. When we hear a tall man say "sit," the sonic picture is actually the same as if a short man were saying "set," and yet we "hear" the tall man's "sit" as "sit" and not as "set."[106]

All such feats, which I am happy to say we accomplish without giving it a second thought, are examples of "categorical perception"— that is, the ability to slot a bewildering number of impressions into categories. In principle, the boundary between these categories—for example, that twenty-five milliseconds—constitutes an unconscious intervention akin to the "not" which we looked at in chapter 1 (fig. 1). We could also call it a digitization process: Our brains snip sounds out of the sound picture in the same way as one would snip holes in a piece of paper: a continuous sequence rendered discontinuous.

This concept of categorical perception can of course be generalized to apply to any process that leads to a differentiation between phenomena. If we were to follow Stjernfelt's suggestion we could even make it general enough to cover the kind of categorization a receptor performs when—with a punctiliousness that puts the border guards of the former East Germany in the shade—it rejects all molecules except those in possession of the right pass, in the shape of a specific three-dimensional structure.

Stjernfelt's aim is very like that of this book: to find a balance between two errors—the one error being to deny that human signs have sprung from a natural world and thus denying that there could be anything in nature bearing any resemblance to human signs. The other error is to deny the uniqueness of human existence, the error inherent in considering all of the world as a kind of human being. As Stjernfelt also suggests, the solution to this problem must be to establish a "natural history of signs," in other words to regard signs as a phenomenon that has, in the course of evolution, freed itself more and more from what it signified, or—in the parlance of this book—acquired greater and greater semiotic freedom.

Thus, categorical perception can be viewed as a primitive version of a sign process, a sort of pre-subjective discriminatory skill, upon which it is possible to imagine evolution gradually building up a series of increasingly complex semiotic systems.

At what point in this gradual progression one can start to talk of true triadic signs—which, of course, imply the presence of an interpretative process—is not exactly easy to define. If we go by the theoretical foundation laid in this book there can be no mention of triadic signs until the system is complex enough to contain self-reference in the form of a duality between the analogic and digital representations. But when do we have that?

Here the minimum requirement is, obviously, a cell. But with a multicellular organism it is debatable whether the individual cell really is a sufficiently independent unit. It could, for example, be argued that its DNA is not self-referential since it refers, rather, to the organism as a whole. But that can hardly be the case, since by this time the DNA has in fact been modified and become specialized enough to serve the functions which the cell in question has to perform.[107]

All my instincts tell me that the cell forms the boundary, the lowest level at which it is reasonable to talk of true sign processes. At the receptor level, on the other hand, we are dealing with proto-sign processes, which are covered quite perfectly by the term categorical perception.[108] Let us take a peek inside the cell.

If we stay with the image of the cell being a Manhattan of sorts, with the individual proteins corresponding to cars, then the first thing we have to remember is that Manhattan, all its skyscrapers notwithstanding, is relatively flat. Whereas the cell could, for example, be spherical in form. So if we are going to draw a true comparison we ought really to think in terms of a three-dimensional street system. Manhattan's layout is actually too simple to stand as a decent model for a cell!

Nevertheless, we will stick to this image, for want of anything better. So: In Central Park we have the chromosomes, winding themselves into never-ending loops. If we let the streets around the park—Fifth Avenue, Central Park West—represent the membrane surrounding the nucleus, then a constant stream of RNA signals is being transmitted through the park gates, by dint of which the proteins (the cars) are built just on the other side of these streets. Now comes the big question: How do the proteins know how to get to the right spot, or, to put it another way, how do the cars find their way around the city?

It is only in the last ten years that the answer to this has started to become clear. But if I were to answer this in detail (as with the G protein), I would soon wear out my reader. So I will confine myself to a rough figurative description (giving the biological terms in parentheses).

Every car has a characteristic spatial design which enables the appropriately equipped auxiliary vehicle to identify it. An auxiliary vehicle is a vehicle that guides selected cars to a specific district. So, with the aid of an auxiliary vehicle our car arrives at its predetermined district. Here the car is then fitted with some additional features (e.g., various carbohydrates are attached). Thus marked it is steered, together with other cars, onto large transporters (vesicles) which drive them to new destinations. These transporters run along special highways (microtubules). This line of traffic may proceed through several stages before, at long last, our car reaches its final destination. In due course all cars sustain some damage—a fact which is noted by other vehicles whose job it is to escort damaged cars to the junk yards (lysosomes), where a whole bunch of special trucks crush them into tiny pieces from which new cars can be built.

All of this takes place at a speed that would leave pedestrians in Manhattan shaking in their shoes with fear. And all the while that the proteins are being directed around the city, hordes of signals are also being transmitted from and received by the chromosomes in Central Park.

If this comparison between the cell and Manhattan seems in any way questionable it is not because the cell's transport systems are any less intricate than those of Manhattan; if anything the opposite is the case. But Manhattan is obviously different because—unlike the cell—it is inhabited by intelligent beings. The cell's "intelligence" comes solely from habits formed over the course of its billion-year-long history.

This inherited faculty is, however, sophisticated enough to run a city like Manhattan in three dimensions, while co-ordinating its growth, metabolism, and service output with trillions of other similar urban societies. The two scientists who, in *Scientific American,* credited the cell with the ability to choose cannot be dismissed as glib commentators. I, at any rate, find it hard to see why the cell—this incomparable piece of evolutionary design—should not be regarded as an independent, interpretive entity.

Here, just to be on the safe side, I should perhaps point out that I do, of course, accept that the processes which take place inside the cell are standard causally determined phenomena. But surely the same can be said of the processes taking place right at this minute in the reader's brain as he attempts to make sense of my words. I am assuming that no miraculous phenomena are occurring there either. To the extent that anything is happening—and it goes without saying that I hope there is—I would suppose it to center around changes in patterns of electro-chemical cellular processes distributed throughout most of the brain, patterns which, if I am lucky, have something to do with the distribution of printing ink on this page.

The point is, though, that in both cases we are dealing with processes that are organized according to a form of logic which reflects the system's (the cell's or the brain's) evolved semiotic function. Understanding these systems causally—besides being impracticable because of its astronomical complexity—could never provide us with the revelations we seek. It would be difficult to provide an explanation of the traffic lights at street intersections in Manhattan based on a physical, chemical analysis of the city's electricity supply. What we are looking for is some insight into the practical principles of how the cell or the brain works, i.e., the system's inner logic, which is, as we have seen, an evolutionary product shaped in accordance with the conditions set by statutes at the semiotic level.

Before leaving the innards of the cell we should perhaps take this opportunity to throw a little more light on the roles assigned to the DNA and the cell. In our analogy, the description lodged within the DNA corresponds to the cars in the city. But this description is not good enough to allow a car to be built or find its allotted space without the help of other vehicles. Descriptions of these vehicles can also be found in the DNA lexicon. But these descriptions would still be useless without the prior existence of a city full of cars. And since the DNA itself never budges from its allotted place, but just waits passively to be visited by certain special convoys of cars which can (1) find the right spot on the DNA; (2) make some kind of copy of that spot (RNA); (3) process this copy so that it makes sense; and (4) steer the copy out to the automobile plant on the other side of the East River, then I hardly think it reasonable to credit the DNA with being the conductor inside the cell. The DNA remains a passive script to be read by the cell, or—if you prefer—a musical score, selected passages from which the cell's instruments are capable of playing without any guidance.

Should the reader still be in any doubt as to the cell's ranking as a competent player in semiotic processes, it might help to cast an eye over the most wonderful of all endosemiotic processes, embryogeny.

Embryologists, who make a study of this process, have found that the fate of the individual cell at any given time is to a great extent determined by its individual history.[109] "During the progression of differentiation," writes the French embryologist Rosin Chandebois, "certain properties possessed by cells at any given time will leave behind ineradicable traces which, together with traces of earlier activities, lead to more and more complex structures."[110] While Gerald Edelman says: "Just what reactions will take place when cells interact depends in part on the history of the cells: on what interaction they have had with other cells in the past. What is more, because such interaction depends on what cells surround any given cell, cells react differently in different places."[111] Thus, every cell has a "positional history"—to use an expression defined by Chandebois[112]—a history which stretches back over innumerable generations of cells. We cannot hope to understand the "choices" which the cell makes without some knowledge of this positional history.

Gerald Edelman's group at the Institute of Neuroscience at Rockefeller University has made a study of the surface proteins, CAM (Cell Adhesion Molecules), which are responsible for the semiotic processes carried out between the individual cells in a section of tissue. These proteins have a crucial part to play in deciding how strong the bond between the cells should be and hence it seems very likely that they exercise a powerful influence over many tissue properties. The gene for one of these, N-CAM, has been found to contain a total of eighteen interruptions (known as introns). These eighteen interruptions consist of greater or lesser chunks of non-relevant DNA, inserted between those sequences which do in fact denote N-CAM. The cells have access to mechanisms for removing these irrelevant sequences while splicing the proper sequences together. But in the case of N-CAM this splicing seems to occur in many different ways, by means of which a whole string of mutually exclusive N-CAM molecules can be constructed. And here we have a splendid instrument for modifying tissue properties during ontogeny.

One highly interesting, but often overlooked, factor of ontogeny is cell death. Healthy cells will suddenly take to manufacturing an arsenal of lethal proteins, which the cells then turn on themselves. This results in cells being detached from their neighbors, shrivelling up and finally being broken into fragments that are soon "swallowed up" by other cells.

When this happens during embryogeny it is usually seen as a formative process. In more highly developed vertebrates, for example, a systematic loss of cells occurs on the palms of the hands and soles of the feet as the cells between the phalanges are eliminated. So the five fingers on a hand ought really to be regarded as four spaces. Massive cell death occurs, not least, during the development of the brain. Forty percent of the nerve cells formed early on in the development of a chicken embryo will die off again.[113] In such cases, biologists will actually talk of programmed cell death. How all of this is controlled is just one of many endosemiotic mysteries. Calcium ions are said to act as signals,[114] but that still does not explain a great deal. Why do the cells interpret calcium ions as a signal to commit suicide?

Cell death also has a part to play in fighting infection. Cells that have been infected by a virus are, for example, tracked down by so-called "killer" T cells, one of the immune system's many types of white blood cell. There has been a good deal of debate on the question of whether the T cell actually murders the infected cell or just tricks it into committing suicide, since the victim's own mechanisms must be undamaged for the "killer" T cell to induce its demise.[115]

The manner in which the T cell tracks down the infected cell is, in itself, an intriguing story. Many viruses, including the HIV virus, have developed the knack of hiding inside cells so that, initially at least, the infected organism's antibodies are not activated. Usually, though, the organism does spot the infected cells, since the latter very kindly break down the virus' proteins and "present" fragments of these on their surfaces. This trick makes them easy for the "killer" to spot, after which they are forced to commit suicide.

This brings us straight to the heart of the organism's innermost semiotic network, the so-called immune system—i.e., the complex of cells, white blood corpuscles, that are carried through the organism in the bloodstream and lymph. Read anything about this system and one will almost always be told that its job is to defend the organism from infection and harmful alien substances which have somehow managed to enter the body, or to defend it against the body's own damaged cells, including any which may have developed precancerous tendencies. So, by this way of thinking, the immune system is a kind of Department for Undesirable Aliens which, with all the zeal for which such bodies are renowned, neutralizes every micro-organism or chemical compound effecting illegal entry.

Since the immune system recognizes everything that is foreign to the organism it must, in some sense, be capable of recognizing the organism itself. The immune system defines a "biochemical self" inasmuch as it reacts to everything that is not "self." This negative definition of the self as something that can *not* be "seen" by the immune system makes immunology a rather odd subject with which to work. After all, doesn't it seem strange that one's own cells should be termed antibodies while the foreign bodies which the antibodies are chasing are allowed to go by their own name?

In accordance with this viewpoint we talk of illness on those occasions when, in spite of everything, the immune system reacts to the body's own proteins. Hence, diseases such as rheumatism, sclerosis, and insulin-dependent diabetes are referred to as autoimmune diseases, i.e., diseases arising from a body having developed an immunity to elements within itself, so that a silent war is constantly being waged in vulnerable areas of that body.

But how does the immune system learn to differentiate between self and not-self? And why do things sometimes go wrong? As yet we have no satisfactory answer to these questions, and there are quite a few paradoxes associated with immunology.

Very briefly, the immune system's astonishing ability to recognize and neutralize alien substances is derived from the so-called lymphocytes' ability to produce an astronomical number of different surface receptors. In his Nobel Prize lecture in 1984, the Danish immunologist Niels K. Jerne put the number of cells (lymphocytes) at 10^{12}, that is to say, ten times as many cells as there are in our nervous system.[116] Every single individual, in fact, every single pre-lymphocyte stage, has the potential to produce more than a billion different receptors of the type capable of recognizing antigens[117] (antigens being the general term for those molecules which stimulate an immune response).

We can skip the biochemical basis for this talent for ringing the changes, remarking only that the variations are presumed to work on the principle of pure chance. Basically, receptors designed to recognize every possible, and impossible, antigen are produced. The question is therefore: How are the cells with the relevant receptors selected and how are the cells with harmful receptors—ones, that is, that can recognize the body's own molecules—eliminated?

The lymphocytes which produce the antibodies are known as B cells. If, for example, a bacterial toxin gets into the bloodstream, there will always be a fair number of B cells carrying just the right receptor for this special molecule on their surfaces. Such a B cell will immediately ingest the toxin and break it down into fragments which are then transported up to the surface of the cell where they are "presented" to the T cells. A tiny fraction of the T cells will possess the exact specificity for the fragments thus presented and these "helper T cells" will then secrete a so-called peptidic hormone, interleukin, which is interpreted by the B cells as a signal to start splitting up. The subsidiary cells, which will be identical to the parent cell, will also recognize the toxin and break it down, and so the sequence is repeated. Thus, before too long, countless cells with the ability to recognize this particular toxin will have been generated.

The B cell antibodies would, apart from one minor detail, be identical to the receptors and, like them, they bind themselves to one distinct section of the toxin, thus neutralizing it. Finally, a few B cells equipped with memories are produced, and these will hold themselves in readiness for a long time to come, in case the same toxin should dare to enter the body again. Immunity has been established. In this elegant fashion, cells with useful receptors are singled out and "filed away."

The question of how harmful cells with receptors aimed at the body's own proteins are got rid of has proved much tougher. The consensus of opinion on this point is that the dangerous T lymphocytes, which are capable of recognizing chunks of the organism's own proteins, are destroyed at earlier stages of their development into immunoactive lymphocytes. This development takes place in the thymus and the theory has it that, while still in the thymus, the immature lymphocytes are exposed to more or less all of the proteins in the body. All that is then required is a mechanism for exterminating those cells whose receptors, even at this stage of development, are being bonded to the body's own proteins. Experiments would appear to have substantiated the existence of such an "exterminator" mechanism.[118] The actual cell-death process is, as it happens, quite similar to the aforementioned programmed cell death.

I admit that the above explanation could be described as a bargain-basement rundown on the amazing world of the immune system. But the reader must remember that my prime concern in this chapter has been to clear a path from the exosemiotic side to the endosemiotic and back again. This path must inevitably run through the sensory apparatus and the nervous system, but it also has to take account of the body washing around the nervous system. There are many indications that it is the immune system which ensures that the nervous system and the body are pretty much inseparable. So we will have to stay with the immune system for a while yet.

It seems likely that the immune system is not in fact geared primarily toward the "non-self." It seems far more likely that it is engaged in the perpetual task of *defining* "the self" at the biological level.[119] In the light of immunology's origins in the fight against infectious diseases, the

"Department for Undesirable Aliens model" is understandable. But it looks very much as if, nowadays, it is merely forcing us to adhere to an unnecessarily distorted point of view.

In 1976 Niels Jerne made an important observation. He pointed out that some antibody molecules are regarded as "non-self" by the very organism which has created them.[120] This leads to the production of anti-antibodies—also known as anti-idiotypic antibodies. In some cases this anti-antibody resembles the original antigen. In other cases it will be quite different. But if it is different, anti-anti-antibodies can be produced, or even anti-anti-anti-antibodies!

And now the reader must be starting to discern the outline of a new self-referential hall of mirrors, *the immunological network*. "In its dynamic state," writes Jerne, "our immune system is primarily self-centered, inasmuch as it produces anti-idiotypic antibodies for its own antibodies, which constitute an overwhelming majority of the antibodies found in the body."[121]

In order to make sense of all this it may again be necessary to consider the question of multicellularity and individuality. It might well be that, ordinarily, we are somewhat dazzled by the feeling of unity our consciousness is capable of producing. We feel like one whole person. And, bearing in mind what a wondrous work the body is, this feeling is, admittedly, well-founded. All of our trillions of cells seem to work together in quite perfect harmony, so perfect that many of us can be well up in years before all the little faults and failings truly start to pile up. And yet one could be forgiven for asking whether there might not still be the odd slip-up. This feeling of unity within ourselves is after all something created in our heads and, as we shall later see, is something of an illusion. How painless a symbiosis do the cells actually enjoy within the bodily community?

As usual, Lynn Margulis and Dorion Sagan have some hard-hitting words to say on this score: "Real organisms are like cities: Los Angeles and Paris can be identified by their names, by their city limits, and by the general lifestyles of their inhabitants. But closer inspection reveals that the city itself is composed of immigrants from all over the globe, of neighborhoods, of criminals, philanthropists, alley cats, and pigeons. Like cities, individual organisms are not Platonic forms with definite borders. They are cumulative beings with subsections and amorphous tendencies."[122]

The biologist Leo Buss introduced the expression "somatic ecology" to describe the bodily dynamic that regulates any potential conflicts between the cell and the individual.[123] And in all probability it is to this we must turn to discover the purpose of the immune system. Its main task is, quite simply, to tend this physical "ecology"—the job of establishing a defense against infection being just a natural offshoot of this task. Taking Buss' ideas a stage further, Francisco Varela calls the immune system "a self-referential, positive assertion of a coherent unity."[124]

The full import of this revision of the immune system's function does not become clear until we ask ourselves an even deeper question: Is there in fact any such thing as an immune system? This is obviously a rhetorical question. Since nothing in the natural world can be isolated from the rest of nature, the boundaries of all natural systems are indeterminate. Nonetheless, every profession creates its own systems and just as the ecologists have devised ecosystems for themselves, so the immunologists have created immune systems. As one historian so aptly put it: "The immune system was a kind of metaphor. It solved the need for communication not only between cells but also between the professional immunologists."[125]

Since all of these systems are constructs, it ought also to be quite legitimate for us to do away with them if they start to have an unhealthily ossifying effect on our thinking. The reason the idea of abolishing the immune system is now being mooted is that its discreteness from the rest of the body and, not least, the brain is beginning to seem more and more illusory. There are many who now believe that the immune system's 10^{12} cells actually form a *floating brain*.

A high level of intercommunication exists between the nerve cells and the cells from the immune system. In the first place it has been shown that not only do nerve fibers run to muscles, blood vessels, and glands, the nerves also branch out into the organs of the immune system—the thymus, lymph glands, bone marrow, and spleen. But the one thing, above all else, that erases the dividing line between these two systems must surely be the discovery that the receptors—which, until now, have been regarded as a characteristic feature of the central nervous system—are also found in the mobile cells of the immune system.

These particular receptors recognize and respond to what are known as *neuropeptides,* hormones produced in the central nervous system. A fair number of these neuropeptides have now been identified and in most cases they have an effect on our behavior and temperament. That they also have an effect on the immune system is indicated by the fact that the monocytes—one of the many cell types in the immune system—will actively gravitate toward the source of these neuropeptides (chemotaxis),[126] and that they are capable of altering the sensitivity of other lymphocytes to antigenic effects.

"Neuropeptides and their receptors . . . thus join the brain, glands and immune system in a network of communication between brain and body, probably representing the biochemical substrate of emotions," as Candace Pert and her colleagues put it in 1985.[127]

What we are dealing with here is two-way communication, since the immune system hormones also exert an influence, through the receptors, on secondary functions in the nervous system. The body temperature, for instance—which is regulated by the brain centers—is highly susceptible to any changes in the level of activity in the immune system. Not only that, but cells from the immune system work their way into the brain, which everyone was wont to consider as being protected from that kind of intrusion by the blood-brain barrier. In light of this fact, Pert and her colleagues suggested that macrophages (monocytes at a later stage of development) act as a "mobile synapse" taking information from one part of the body to another.[128]

Candace Pert is not one for keeping her ideas quiet. In 1988 she told the magazine *Woman in Power:* "There is no mind/body, controller/controlled, male/female, neuron/Glial cell dichotomy. Rather there is a 'mindbody-bodymind,' a dynamic system."[129]

If the inside of the brain is in fact awash with volatile, immune-system cells and hence not merely linked to the body by the nerve impulses, then we are looking at a far more integrated system. The notion of a centralized government in the brain fades away, to be replaced by a more interactive and analogous organization.

Thus the immune system becomes the mobile extension of the brain within the body, engaged in a constant race around redefinition's hall of mirrors: Feelings reflected in chemistry, chemistry reflected in chemistry, chemistry reflected in feelings, feelings reflected in feelings.

On the triadic ascendance of dualism

Again I feel my sangfroid being shaken by the critical reader. It is not so much that I am afraid she is going to refute everything I have said up until now, even though she may well find it a bit obscure in parts. It is more that I suspect her, deep down, of thinking: "Okay, so what? Isn't that just the point, that we have a well-oiled biological machine to thank for all of this insight, and doesn't this just go to show that biology is doing very nicely, thank-you, and has no need to revise its basic principles? To put it bluntly, what do we need all your bewildering sign processes for when everything is running so smoothly on orthodox cause and effect models?"

In reply, it ought first to be said that it is an illusion to believe that there is any such thing as a natural science in the sense of a science dealing with nature, or a biological science in the sense of a science dealing with living things. There is no one science that takes nature, let alone living things, as its subject. Biology is a loosely constructed network of fifty or more different sciences which only deserve to come under a common heading because, on the whole, whatever they have in common is arguably greater than anything they might have in common with adjacent sciences such as physics and chemistry on the one side, or psychology, anthropology, and ethics on the other.[130] It can be a far cry from a biologist sitting in a museum studying the familial relationships of a group of fishes to a biologist sitting in a hospital studying the passage of macrophages across the blood-brain barrier.

It might be argued that biologists are united by one common and fundamental theory, the theory of evolution, which acts as a sort of umbrella under which all of the individual sciences can be lumped together into one body. But why then do psychology and anthropology not come under the biology umbrella? After all, human beings are also a product of evolution.

In actual fact, what biologists work with is not living things but *data*. Listening to the current rhetoric in scientific circles one could easily be misled into believing that data is something hidden within the natural world, something which the good experimentalist goes out and most cunningly coaxes out of it. But data is, of course, by no means part of the natural world. Right now, for instance, my watch gives the time as being 11.44.06. That is data. But, for obvious reasons, these six digits are not in themselves a part of nature—or what we would normally understand by that word. For one thing, time itself does not consist of digits, much less seconds, does it? And for another the digits are "wrong" because my watch is running approximately half a minute fast. If, however, all the people in the world agreed to push 00.00.00 hours half a minute forward, then the data provided would be correct. So correct answers are arrived at by agreements reached between people. And so on and so forth.

That data does not speak for itself, that it is what in scientific jargon is referred to as "theory loaded"—i.e., it only makes sense given the prior acceptance of the idea of an impenetrable system of theoretical premises—seems now, after fifty years of challenging positivism, to be verging on common knowledge. But possibly nothing demonstrates more clearly how little natural science has to do with nature than the fact that, for the most part, scientists could not care less about such "philosophical jiggery-pokery." Questioning the nature of life comes under the heading of philosophy. To the scientists, reality amounts to data plus those theories which make sense of the gathered data. So questioning the connection between data and reality means betraying, right at the start, that one is not a true scientist.

I make no secret of the fact that it pained me to discover this was the case. Back in the seventies when, as a young biochemist, I first realized how serious the problems of pollution and global ecology were I began for the first time to suspect that a serious imbalance existed between living things and the science of living things, that is to say biology.

But the environmental crisis was far from being the only disturbing factor. The interminable debate on nature and nurture, or the recurring assertions as to the genetic defectiveness of deviants (including women, other races, criminals, and schizophrenics) revealed that ideological notions were alive and kicking within the field of biology. The determination of humanity's place in the natural scheme of things was far from being in safe hands. This led to a suspicion that even the theory of evolution might be flawed. How reliable, really, was neo-Darwinism's belief in natural selection as a magical formula for explaining the wonders of Nature? How could one, on the basis of data drawn from the goings-on in boxes of fruit flies, extrapolate on the billion-year history of the living world?

In my naïveté I felt, as a *natural* scientist, increasingly compelled to investigate all of these questions. And only after spending fifteen years on this, did it dawn on me that I had been barking up the wrong tree. I was now being regarded as some sort of philosopher, something which the philosophers—quite rightly—would have none of. While still intent on understanding Nature I had allowed my search to focus on misapprehensions, and this had forced me to bypass the concrete data that constitute the traditional mainstay of science.

What does this show us? Well it must at least suggest that the links between science and nature are by no means unassailable and should be subject to constant checking. On the face of it one could be forgiven for believing that the stream of technological triumphs with which science and medicine enrich society would be enough of a guarantee in themselves. If natural science had nothing to do with nature how could it possibly work? How could medicines relieve pain, or the gene-splicing of bacteria cells produce insulin?

But the question is just this: How well does it in fact work? Take a look at modern Western society. Are not all of our greatest problems related to the interface between nature and culture, those areas where the interplay between nature and culture is most intense? Just when scientists are about to map out human genetic material *in toto,* any responsible person is bound to fear that the absence of a proper understanding of human beings as bio-psycho-social beings will turn such knowledge into a threat. And our sense of wonder at the surgeon's ability to perform heart transplantations is rather laughable if one considers that we are still absolutely at a loss to explain the placebo effect: The common calcium tablet—a placebo—actually seems to be the most effective, all-round medicament we have come up with to date. No one knows why. I have already mentioned the perpetual in-fighting between technology and nature. We have no idea what to think about the various consequences for humanity of biomedical technology. We have no real conception of how humanity can have evolved out of a mindless natural world and we have not even the beginnings of a decent theory as to what consciousness actually is.[131]

In short, we have not even touched on the question of *humanity's place in nature*. How could we when we have consistently placed people—as minded beings—outside the brackets in the study of natural history? Which reminds me of the words of the Danish philosopher Hans Fink: "Anyone who places himself outside of Nature will feel out of place."[132]

I must, therefore, answer my critical reader by saying that everything is running far less "smoothly" than she might think. We are very badly in need of ideas for improving the biological scientific models of the course of life on Earth. Despite the impressive volume of data which biology and medicine can produce, it is impossible to rid oneself of the suspicion that there is a chronic gap in all the information they keep churning out—a particular type of non-knowledge as I called it in an earlier book, *Naturen i Hovedet (Nature in the Mind's Eye)*.[133] A non-knowledge which leaves these sciences incapable of supplying solutions to the inflamed relationship vis-à-vis nature into which they themselves, thanks to technological spin-offs, have put culture.

The problem is that, while Descartes' dualism may have been ridiculed, it has never been totally uprooted. Unbeknown to those scientists who perhaps are not even sure themselves what dualism is, Descartes' old theory is still controlling science's understanding of its own limitations. In the original, Cartesian version the dualist theory considers the world to consist of two fundamentally different types of matter: the ordinary physical matter—res extensa—which has material form and can thus be seen and sensed, and then the substance of thought or—in more general terms—of the mind, res cogitans, which has no physical, sensory manifestation. According to Descartes res cogitans was exclusive to human beings; in his eyes, animals were nothing but machines. And the interaction between the intellectual and the physical elements, opined Descartes, was effected by a little gland in the brain known as the pineal gland.

The idea of such a "rational substance" has not attracted many followers in recent years. Ask any of today's neurobiologists or neuro-psychologists and most of them will reject such a notion as absurd, and instead declare themselves to be materialists. That is to say, they work from the premise that there is only one kind of matter and energy, con-sistent with what is recognized by modern physics. As far as they are concerned, that machine up there in the skull needs no "ghosts" to get it going.

On that last point I would have to agree with them. But what the materialists seem to overlook is that the idea of a rational substance has left a pernicious mark: The psyche and the body are still regarded as either-or phenomena. And since this rational substance is unacceptable all we are left with is the purely physical. What gets left out of such an analysis is the evolutionary perspective, the idea that a system could be *more or less rational*; that rationality is something that can occur at levels other than that of the human psyche. If reason does not spring from some special substance but is inherent, rather, as a potential within the physical material—a potential with its ultimate roots in those "lumps in nothingness" that we spoke of in chapter 1, and which a forgetful (mortal) self-referential system might manage to "tell" its descendants about—then this opens up a non-dualistic perspective that has never been tested simply because it conflicts with the dualistic either-or philosophy which not even materialism seems able to shake off.

To quote John Searle: "In denying the dualist's claim that there are two kinds of substances in the world or denying the property dualist's claim that there are two kind of properties in the world, materialism inadvertently accepts the categories and the vocabulary of dualism. It accepts the terms in which Descartes set the debate. It accepts, in short, the idea that the vocabulary of the mental and the physical, of material and immaterial, of mind and body, is perfectly adequate as it stands. It accepts the idea that if we think consciousness exists we are accepting dualism." Thus, materialism presupposes dualism unless, that is, it denies the existence of consciousness. Searle then goes on to suggest that "if one had to describe the deepest motivation for materialism, one might say that it is simply a terror of consciousness."[134]

To modern science, dualism still holds good as a way of dividing the world into two kingdoms, those of mind and of matter, the cultural and the natural spheres. Non-intervention is still the easiest compromise and one which ensures that both the humanities and the natural sciences can get on with their work undisturbed. And it is this boundary that biosemiotics seeks to cross in hopes of establishing a link between the two alienated sides of our existence—to give humanity its place in nature.

By insisting that life must be seen as a network of sign processes the biosemiotic view automatically leads us away from the standard "chain of command models" of the brain-controlling-body or DNA-controls-embryogeny type. The whole essence of the sign process is that the decentralized units at tissue or cell level can interpret their own environment and act accordingly. While theorizing based on causality, more or less as a matter of course, assigns an authoritative body to the model, from which these unilateral decisions emanate, semiotics paves the way for "self-organizing chaos."

Here I have enlarged a little upon the expression "organized chaos" which the Danish biologist Axel Michelsen uses to describe the way in which a colony of bees functions: "The only situation in which the queen can perhaps play a decisive part is during the swarm's flight to a new hive," writes Michelsen. "As we have seen, the selection of a new home is left to a handful of scout bees. This is, however, the only known instance of a decision being made by a small group of bees. In all other activities it seems to be up to each individual bee to make its own decision."[135] In other words, without any governing body of any kind and without being in possession of intelligence tens of thousands of bees—each of whom has a lifespan of just four weeks—manage to coordinate the collection of food, care of the larvae, feeding the queen, building cells, defending the hive along with cleaning and storekeeping duties. "We now understand the crucial role which dancing plays in regulating nectar collection to suit the hive's capacity for taking in food," continues Michelsen. "The most economical course (as regards the deployment of the hive's workforce) must be for food gathering and storekeeping to balance one another out. If the capacity for receiving the nectar into the hive is greater, proportionately, than the number of foraging bees coming back with supplies, the waiting time is short. So the forager bees dance, thus recruiting more foragers. Conversely, dancing and recruitment levels will drop when the storekeepers cannot keep up and the waiting time becomes too long."[136]

The individual bee is not, in our sense of the word, intelligent. It is easily fooled and cannot deviate from its inbred repertoire of procedures. But it is evidently capable of interpreting situations and acting accordingly, within this repertoire. Rather than comparing the bee to an individual we ought to compare it to a cluster of cells, a piece of tissue. Expanding on that line of thought, a colony of bees could be likened to a diffuse organism, its individual sections of tissue crawling all over one another or flying out into the world to collect food and bring it back, so that the organism as such can stay safe at home! The idea of a colony of bees as an organism dates all the way back to Aristotle.

If we then transfer this image back to ourselves or other multicellular but integrated organisms, we might hit upon the particular aspect of our biology that I am aiming at here—autonomous chaos. Or, to use the metaphor that keeps cropping up in this book: Inside us there is "someone." Life is based on the principle of "someone" inside "someone" inside "someone" inside. . . . What emerges, when the authority for interpreting and making decisions is delegated to organs, tissues, and cells, is a hierarchical network of sign processes the accumulated output of which constitutes the coordinated actions of the organism. No single body controls this autonomous chaos, the efficacy of which can only be explained by its actual history throughout all the various stages of discoveries and conquests made by other life-forms. In chapter 9 we will see how not even the mammalian brain can claim the rank of general manager in this store: It is not the brain that does the thinking in a human being, it is not even the body, but the natural history whose children we all are and in which we all have a part to play.

Thus biosemiotics releases the genie of reason from the well-guarded bottle which we know as the human brain and accords it an immanent position in the natural history fairy tale. This move enables us to unite the two separate spheres: Cultural history runs parallel to natural history; at one time they were one and the same.

This line of reasoning can be roughly summed up in the following model which I have recently been using to organize my work:

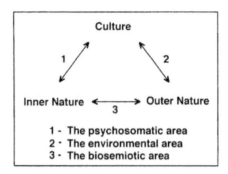

FIGURE 12. *The lost connection: The biosemiotic area as mediator between man's outer and inner natures and, hence, between culture and nature.*

The postulate made in figure 12 states, very briefly, that in order to solve the problems associated with the environmental and psychosomatic areas (arrows 1 and 2) we must learn to understand how we as people fit into the natural world that produced us. In the last three chapters I will be attempting to outline the way in which the bio-semiotic view provides us with a departure point for this task, one which can of course also be formulated as the outline of a bio-anthropological theory. First of all, we need to consider human beings as intelligent, talking animals, then we have to take a look at psychosomatic awareness, and finally we must examine our ability to put ourselves in someone else's place—ethics as a condition of existence.

On language: Existential bioanthropology

I have observed that many of my students—who are, of course, studying to become biologists—are extremely reluctant to accept the idea of human language as something special. They point out that animals such as dogs, whales or chimpanzees might well have a language that we human beings have just been too highfalutin to acknowledge. And, in my experience, my holding up of the novels of Dostoyevsky or the Bible as examples of how human language is something quite unique in this world seems to make no great impact. If anything, I have the feeling that these students look upon humanity, and human nature in general, as a warped work of nature epitomized by its destructive penchant for building concrete blocks of slums and waging war. And, seen in that light, the fact that a few sensitive individuals might find it in themselves to write emotionally harrowing works of fiction does not seem all that strange.

Generalizations are always a bad thing and obviously a great many students have quite different views on the subject. Nevertheless, we have to make a stand against this lack of respect for human language. The thesis which states that language is bad news for nature as a whole and people in particular is most definitely a moot point. While the most intimate, most profound of human experiences can, at all events, hardly be said to be of a linguistic nature, to maintain that language is nothing special is quite absurd. For good or ill language accords us human beings a totally unique position within the natural world. The fact that we have, perhaps, abused this position[137] is no good reason for denying this. And if we ever want to reach an understanding of humanity's place in nature then we will have to account for the emergence of a talking animal and for what having the power of speech means to us.

But language leads us straight into a number of paradoxes. As the Swedish human geographer Gunnar Olsson put it: "Language is both a problem and a resource, since it has to use itself to understand itself."[138] And, in the same vein, Dane theologian K. E. Løgstrup has written: "We can never get beyond language, we are pledged to it. We speak in its favor."[139] Which poses a problem, because the very fact that we acquire the power of speech quite spontaneously "proves that not only is language an armory at our disposal, it is also a power invested in us, it is our mainstay."[140] Language has, as it were, its own independent existence, almost as if it, too, had something on its mind. No wonder philosopher Ludwig Wittgenstein could write: "Philosophy is a bewitchment of our intelligence by means of language."[141]

The idea of animals having their own language is no doubt boosted by an image of language as primarily lexical. Words are seen as being tags of a sort, referring to given phenomena in the outside world. Such a (semantic) system need not have anything to do with the voice; in principle it could be linked to all of the body's other communication systems, to colors, scents, etc. Such a system would at least allow for the possibility of utterances along the lines of "me Tarzan, you Jane." Language at that level can actually be taught to a chimpanzee which has already learned to use sign language or to communicate via flash cards. But, for the most part, gorillas and chimpanzees never get beyond this stage. At the age of two, however, our own children almost explode into experimenting and juggling with language as they test the "power" of words and phrases and their own ability to reconstruct every possible— and impossible—situation of emotional significance to them.

As we shall see, the naïve lexical image of language takes little or no account of the infinite reserves of creativity which are not only a prerequisite for language but to which language gives us access. How could one, for example, within the confines of this image, account for the following sentence: "The ham sandwich at table five is growing impatient"? Now anyone familiar with fast-food restaurants will have no difficulty in understanding that sentence, since one simply has to imagine that this is one waitress's way of reminding another that the impatient customer at table number five still has not been served with his ham sandwich. But the words themselves make no sense unless one believes that a sandwich can be impatient.

So language presupposes that mental or cognitive models of reality can be created "in the mind's eye." Such models are the product of experience and can easily become outdated as witness the following example:

A landlady, about to rent a room to a young lady, tells her prospective lodger: "Now, I won't have men in the rooms." To which the girl replies: "Well that just goes to show, everybody's different. *I'm* all for it."[142]

Language is like a living creature, always changing. And one of the problems with growing older is that it is difficult to keep pace with language. We grow lonely from being misunderstood or perhaps sniggered at even though the things we say are (i.e., were) perfectly obvious and "natural." If, for example, during a lecture, I started spouting some of the weird "hippie" phrases that were common coin among young Danes in the sixties and seventies in an embarrassing attempt to seem young and trendy, my students would probably stare blankly at me.

In citing these inadequate examples I am seeking merely, in the first instance, to persuade the unprepared reader to drop any idea of language as a kind of inner lexicon and venture out with me into the deep waters where language had its beginnings. As we shall see, "the inner lexicon" seems to be a consequence of language rather than a precondition for it.

Just to be on the safe side, we had better begin this attempt to trace the natural history of language by reminding ourselves that none of the vast number of native languages spoken on Earth today can, in any way that is pertinent, be deemed to be simpler or more highly developed than any others—a fact of which even nineteenth-century philologists were well aware. In other words, to find "undeveloped" language we will have to go back to prehistoric times.

Not that we have to go that far back—an amazingly short space of time, in fact. In evolutionary terms, 200,000 years is nothing to speak of and yet we can say, with a very great degree of certainty, that the spoken word is no older than that, and could quite conceivably be much younger—perhaps more like approximately 100,000 years old.[143] For speech to be possible necessitates a larynx structure first found, for certain, in modern man. And the 200,000-year limit corresponds with what appears—going by the genetic data—to constitute the approximate point in time when modern man (and that includes Neanderthal man) first appeared on the scene. The earliest relics of our own species *Homo sapiens sapiens* are, however, only around 125,000 years old.

The extent to which Neanderthal man may have been able to speak in the same way as members of our own species is debatable. But that question is not so crucial to the matter in hand here since the most remarkable feature of this development is, in any case, something quite different: namely, that more or less all of the four to six million years of development leading up to modern man's large brain has come to an end by the time the spoken word manifests itself.

At any rate, in existence four million years ago were a number of apes, the so-called *australopithecines,* who walked upright and whose craniums were somewhat larger (450 cm^3) than those of present-day anthropoid apes. Stone implements first appear around two million years ago with a creature known as *Homo habilis,* who had a cranium measuring 750 cm^3. *Homo habilis* appears also to have possessed some ability to predict and to plan ahead, since stone implements have been found at camp sites situated several kilometers away from the lava deposits from which these stones came. With *Homo erectus'* appearance on the scene one and a half million years ago brain capacity has expanded to 1,300 cm^3, a measurement that was only marginally smaller than *Homo sapiens'* present average of 1,500 cm^3. From *erectus* a slow process of development brings us to modern man's almost shocking emergence, possibly 200,000 years ago, with a radically different cranium formation which allows both for speech and for a wide variety of facial expressions.

On the basis of what we have gleaned from various sources on *Homo erectus,*[144] we are bound to conclude that only when our forefathers had tripled their brain capacity and developed cultural resources conducive to a thriving cottage industry in stone implements; organized hunting in groups, settlements that changed with the seasons; and the use of fire—only then did language develop. Such a conclusion would however be seriously at odds with modern philology and cognitive research. Linguist Thomas A. Sebeok goes so far as to state that even *Homo habilis* must have had command of language since they operated with considerable—and ever-increasing—advance planning: "The point here is that long-range foresight presupposes language, so that, judging by the technological complexity of *homo habilis*, we can safely assume that language must have evolved prior to two million years ago."[145]

But the riddle is solved the minute one simply accepts that the spoken word and language are not one and the same. As John Lyons says: "The fact that we can speak what is written down and write down what is spoken demonstrates the independence of language and speech. It demonstrates that, as far as their verbal component is concerned . . . languages are independent of the physical, or psychological, medium in which they are manifest."[146] According to Lyons, language is made up of many layers which have not "necessarily evolved at the same time . . . or in the same way."[147] He calls this the hypothesis of polygenesis, a hypothesis which suggests that the development of the spoken word, with the accompanying linguistic functions situated in the left (dominant) side of the human brain, should merely be regarded as one last shoot on the tall tree of language.

In chapter 1, we looked at Gregory Bateson's refreshing theory that the roots of language lay in the distance inherent in the "not" concept; a distance which may already have been present, in a proto-version, in the animal's "snap" (which, as we saw, contains the message "this is *not* a bite"). Implicit, also, in this theory is the idea that language has not—at least to begin with—served any communicative purpose (similar in style to that of body language) whatsoever, but that it has more likely been associated with the development of a quite new type of inner, mental concept—let us call it a cognitive model. Not until much later, and after many a winding detour, did these cognitive models give rise to the development of a communicative language in the form of a grammatical series of sounds, words and sentences. Such theories fit in nicely with Lyons's polygenesis hypothesis.

As Thomas Sebeok pointed out, in the light of Jakob von Uexküll's *umwelt* theory these cognitive models can be seen as elements in the specific human *umwelt* (cf. chapter 5): "The models at the disposal of other animals are relatively rigid, with only a few, relatively simple rules. . . . The human model, by contrast, includes not only the objects and events of everyday life ("reality"), but an extremely elaborate and powerful set of rules capable of working radical transformations to suit human purposes and goals. Such a set of rules is called a syntax."[148]

This switch from an *umwelt* containing very few transformation rules to a grammatical *umwelt* is perhaps best compared to the switch from the flash card principle to the Lego principle. While the sunrise depicted on a flash card will always remain the same, able at best to be turned one way or another, pulled out or hidden away, the sentence "the sun is rising" can be played around with in any number of ways. It can for example be split up and reassembled: "the sun" could be replaced by "the bridge" or even "the bread," or "rising" could be replaced by "setting." Some wit might even take it into their head to change "sun" to "son" and so on.

"Humans have evolved a way of modeling *their* universe in a way that not only echoes "what is out there" but which can, additionally, dream up a potentially infinite number of *possible worlds.*"[149]

In the research into the evolution of mankind two schools have been particularly predominant. One of these schools has endeavored to reconstruct the history of evolution on the basis of archaeological finds—on artifacts, tools and utensils for the most part, but also other vestiges of prehistoric civilization. The other has examined the functional anatomy of our forefathers (e.g., the size of the brain or structure of the larynx). Recently, however, the Canadian psychologist Merlin Donald has attempted, in his book *Origins of the Modern Mind*,[150] to bridge the gap between these two schools, with the assertion that, for obvious reasons, the cognitive area represents the meeting point for culture and the mind or, if you like, the source from which language sprang: "Language is a reflection of the intelligence of the speaker; it is limited by the memory, knowledge and ability of the speaker. It receives its meaning from the broader cognitive realm of experience and context. . . . the debate over the emergence of language is also a debate over the emergence of all uniquely human styles of representation."[151]

So—if I may be allowed to employ a well-worn metaphor—the spoken word is merely the visible (audible!) tip of an iceberg, the substance of which is language itself, seen as part of the cognitive system. Or perhaps another metaphor, just as well-worn, would be even better: Speech is simply the drop that causes the cup of language to overflow. Confusing speech with language is little better than calling an automobile a beep-beep.

Merlin Donald's original thesis states that the development of language occurred in two main phases with a quite unique culture evolving—as a halfway house along the route from ape to human—around a mimicked memory, a social form of memory founded on the ability to carry out collective motor-based reconstructions of earlier incidents. This *mimetic culture*—which succeeded in spreading from Africa to Asia and Europe where it prevailed until just a few hundred thousand years ago—was reached with *Homo erectus* one-and-a-half million years ago and provided the necessary cognitive precondition for the subsequent development of actual speech in *Homo sapiens.*

The fact that the power of speech is not just some quirky detail that sets human beings apart from anthropoid apes can be seen, for example, in the chimpanzee's aptitude for play. Our own children will get caught up, quite spontaneously, in any number of new games simply by imitating other children. They soon pick up the rules, often without help. In most informal games, words play very little or no part and deaf and dumb children have no trouble in learning the rules. And yet even these games are quite beyond anthropoid apes.[152]

Another example, and one which I believe touches upon the whole crux of bioanthropology, is our children's ability to point. Children start pointing at things they want at around fourteen months old. Although the significance of this pointing may seem very obvious to us, it does in fact entail an exceptionally complex mental process. And most of us know how a dog will just stand there quivering with expectation and gazing dumbly at that extended forefinger. During the stage immediately preceding the child's mastering of the pointing ploy, it acquires the far-from-insignificant knack of directing its gaze at that point in the room on which its mother's eyes are fixed. In order to do this the child has to plot, as it were, the course of its mother's gaze, a phenomenon which, in physics, is denoted by the term "a vector." Not only that, but the child then has to figure out what the specific object of its mother's attention might be.

What this shows us, therefore, is that the act of pointing is not simply a spatial skill, since it presupposes the child's ability *to perceive an intention behind its mother's gaze.* Chimpanzees are incapable of performing this feat; they lack, writes Donald, "the ability to produce conscious, self-initiated, representational acts that are intentional but not linguistic."[153] Human beings alone have the ability to identify with others, which also explains the point made earlier that no one has ever reported having seen chimpanzees comforting one another (care and nurture of young having nothing to do with comfort in this sense).

A vital key to the understanding of the cognitive abilities of anthropoid apes is provided by a phenomenon which has been dubbed "social intelligence." Roughly, social intelligence is the ability to deal with social complexity. Animals living within a complex group-based social system must be capable of understanding and remembering the vast number of dyadic two-way relations in force among the individual members of the group and must also be fit to cope with the set of rules and customs which govern the life of the group. The relatively large size of the ape brain (a measurement which confounds the theory that bigger animals have bigger brains) has been shown to increase in proportion to the size of the social groups to which they belong.[154] And among the big anthropoid apes the relative brain capacity is twice as large as that of the average mammal. The reason for such a state of affairs is probably that large groups prove to be unstable if every single dyadic relation existing between the animals in the community is not monitored regularly.[155] So considerable pressure is being brought to bear on the animals' capacity to mentally accommodate all of these relations. And in this area the chimpanzees and gorillas are past masters.

For this reason Merlin Donald calls the anthropoid ape's cognitive style, or culture, *episodic:* "Their lives are lived entirely in the present, as a series of concrete episodes, and the highest element in their system of memory representation seems to be at the level of the event representation. Where humans have abstract symbolic memory representations, apes are bound to the concrete situation or episode."[156] The memory of anthropoid apes could also be said to function as a set of previously experienced episodes that can be conjured, in the form of flash-backs, on a mental screen.

Besides the *episodic memory* there is also a so-called *procedural memory.*

This procedural memory provides a more basic complement to the episodic memory. It enables the animal to perform a learned sequence of actions—e.g., catching an object in mid-flight. Thus it is geared toward the general procedure underpinning an action, while the episodic memory actually files away specific actions or situations. But episodic and procedural memory are also dependent on a variety of neural mechanisms, as witness the case of birds who lose their ability to sing (procedural memory) if injured in one center of the brain while injury to another center will lead to the loss of the ability to seek cover (episodic memory). Something similar has been observed in a patient suffering from a dreadful form of amnesia occasioned by neurological surgery. While this patient was perfectly capable of acquiring new motor skills he was absolutely incapable of remembering specific incidents in his life. He had, for instance, to be introduced time and again to the doctors who were treating him, and as far as the skills which he had obviously acquired were concerned he could never recall *having* ever acquired them.[157] His episodic representation system had simply gone AWOL, without this having any effect on his procedural memory.

While all animals must possess some degree of procedural memory, episodic memory appears to be unique to birds and mammals. But human beings, over and above possessing the two above-mentioned types of memory, also have what is known as *semantic memory*[158]—the ability to remember meaningful relations without these being linked to any specific situations.

The anthropoid apes have taken episodic memory as far as it can go. No other animals can handle such complex social situations as well as chimpanzees and gorillas. And yet these animals are still unable to rise above the actual situations. Their intellect does not allow them to distance themselves from situations and break them down into isolated elements with an independent significance distinct from the picture as a whole. And so they cannot, mentally, trigger off situations. While they might well be able to shuffle the various individual elements—the pictures on the cards—around, they are still tied to the hand they have been dealt. Their imagination is, as it were, caught in an unbreakable holism.

And therein lies the key to their poor performance when it comes to learning sign language. In order to remember how to use a sign apes have to consult their episodic recall system, which produces a memory whereby an association is made between the specific stimulus, response and reward once employed to help them memorize the sign. The sign is never truly distanced from the actual situation, it has no real freedom of movement. One sign cannot take another by the hand and stroll of along the open roads of language, because both signs are stuck in their own all-too-concrete situation.

How to loosen these bonds is the key issue in bioanthropology. And by now we should be ready to tackle this process.

It is a fascinating thought that Homo erectus, with a brain capacity not really so far removed from the present dimensions of our own brains, may have been very much like us in almost all respects—and, especially from an emotional point of view, in the most profound ways. Why, were it not for that one little quirk—the lexicon, that ability to send our inner experiences flowing from our lips in streams of words to be pondered and debated among ourselves, adopted or rejected—we might be said to have been almost identical at birth. The difference between the talking Homo sapiens and Homo erectus may not have been any greater than that between the people of the later Stone Age and the people of today. Because what separates modern man from Stone Age man is the existence of external (extrasomatic) memory banks—first and foremost the written word in the form of books, but also the legacy of sculptures, pictures, buildings, tools and, these days, computers. The presence of these external memory banks implies that we, as adults, bear the burden not only of our own inherent intellectual legacy but also of a hundred-generation-long struggle to extract the essence of our forefathers' experience. This struggle has taught us to live in a world saturated by science, technology and art—a world which could quite conceivably create an even greater gap between us and the mind of Stone Age man than Stone Age man, by virtue of the spoken word, created between himself and the mind of Homo erectus.

It could perhaps be argued that there is no comparison between these two mental leaps because the one, the leap from *erectus* to *sapiens* concerns inborn attributes while the other, the leap onwards from modern man relates to learning. And this is, to a limited extent, true enough. But from a biosemiotic point of view they cannot be separated in this way. In his book *Neuronal Man* the French molecular biologist and Nobel prizewinner Jean Pierre Changeux makes the point that the young brain extends its nerve endings in quite haphazard fashion, creating between them, in the process, a vast amount of connections—synapses—which correspond to potential lines of development. Only a very few of these connections will survive, however, due to the ongoing process of cell death[159] (cf. "programmed cell death," chapter 6). This results in the cultural stamp leaving a telling imprint on the neurological structure of the brain. If not at birth then certainly by the time they start school modern human beings are therefore already very different from the people of the Stone Age. Just to be on the safe side it ought also to be mentioned that this restructuring of the neurological terrain is not altogether irrevocable. It appears at any rate that, even in adults, the area of the cerebral cortex which registers hand movements can be expanded or reduced as required.[160] But for anyone desirous of reverting to the Stone Age mentality it would hardly be enough just to journey back to the settlement at Vendsyssel-Thy. You would have to retreat pretty far into the Siberian taiga and stay there for years, and in fact it would be best to start out in early childhood.

My aim, with this little digression, is really just to indicate that even though it may not seem such a far cry from *erectus* to *sapiens* in emotional and neurological terms, a world of difference separates these two, thanks to this one factor—the spoken word. The appealing thing about Donald's idea of *erectus* as a, in many respects, highly developed halfway house between apes and man is the very fact that the development of the spoken word would appear to have been an impossible task for evolution had it not occurred in a creature which was, from a cognitive point of view, already almost human. Because the problem is, as psychologist Jerome Bruner (among others) has pointed out, that speech is a "social skill" which is meaningless without a "social environment."[161] Speech supplies human beings with a quick and efficient method of creating and transmitting symbols. But what good would we derive from such a method if a cognitive community, in which it could be used, did not already exist. Speech demands both a coding mechanism (in the speaker) and a decoding mechanism (in the listener). But before evolution can establish neural structures in the brain, designed to perform these functions, some sort of overlap between the models of the surrounding world which the individual minds in the linguistic environment are busy creating must be established. A collectivization of the individual *umwelts* needs to be achieved.[162] And here the idea of a mimetic culture as a midway point seems just the resource we were looking for.

The first stage in the move toward a mimetic culture was Australopithecus. What it is, in survival terms, that sets this or that creature (for there were many different types) apart from the chimpanzees' ancestors is hard to explain, since an upright posture is a far less efficient method of getting around than going on all fours, and the potential benefits of leaving the hands free[163] do not appear to have been fulfilled by the ability to fashion tools.

And yet *Australopithecus* might still hold the ultimate key to a strategy for survival which appears to epitomize every stage of the human race's development: the social community. To my mind at least the most convincing explanation for the emergence and survival of a creature as odd as *Australopithecus* is provided by C. Owen Lovejoy's theory on the participation of both sexes in the care of their young.[164]

Lovejoy's idea is that the development of a large brain presupposes a prolonged childhood and, hence, an increase in child-rearing duties. *Australopithecus'* way of overcoming this limitation has, according to Lovejoy, been for both sexes to do their part in caring for the young, something unheard-of among chimpanzees. For this strategy to work required a strengthening of the bond between females and males, a more permanent form of dwelling and, not least, the setting up of a communal "cooking pot" to which both sexes would contribute. This kind of "shared meal" is rare among groups of chimpanzees, where individuals are more likely to help themselves as they go along, leaving mothers with the sole responsibility for feeding their offspring. So the chimpanzee could not afford to develop a longer childhood and a bigger brain. But with this scenario *Australopithecus'* upright posture can be accounted for, as an adaptation to meet the need for carrying food back to camp over long distances. Having the arms free must have been a big help here.

The speculative aspect of such theories is something we will just have to live with, there being very little evidence to back them up. But what is so tempting about Lovejoy's theory, in contrast to so many others, is the way it focuses on the social dynamic. As we have seen, this social complexity has actually been found to correlate with the relative brain capacity of mammals. Furthermore, it is hard to believe that the knack of banging stones together or gathering food in the forest should be as great a test of cognitive ability as the need to comport oneself in a complex social context.

Donald defines "mimetic skill"—or *mimesis*—as "the ability to produce conscious, self-initiated, representational acts that are intentional but not linguistic."[165] Thus reflex, instinctive or routine acts are not mimetic. Nor is mimesis merely imitation since it is imitation with a specific purpose—the aim being to reenact or relive earlier events. There may be many psychological and social reasons for such behavior but the main advantage would appear to be three-fold. Firstly, control is gained over the emotional side of an event—without that, reenactment would be impossible. Secondly, control is gained over the individual's own physical scope. The brain must be able to supply the mind with a map of the body and all its potential actions and expressions. One curious example of this is the smile. A baby instinctively smiles at its mother, even though it has never seen its own smile. Only by reference to its own mental map can the child tell what its own smile is. The third benefit is that mimetic reproduction turns events into a collective resource which reinforces the group's sense of emotional unity and—perhaps more than anything else—aids the development of the group's model of itself.

"Charades" has, for generations, been a popular and hilarious party game. A talent for mime and the fun of acting out and trying to interpret a series of mimed actions testifies to how deeply this mode of expression is rooted in our consciousnesses. In Denmark, as in a number of other countries, this game formed the basis of a popular prime-time television series. But my own interest in these shows waned once I realized that the game seemed to depend much more on sign language than on true mime. The performers were making use of a whole range of signs, all with a distinct linguistic connotation.

I imagine that similar conventions must gradually have become established among our *erectus* forefathers. Particular, recurrent incidents which were part of some greater narrative context—like, for instance, the threat "don't do that"—were perpetuated in the form of standard gestures or standardized sounds which everyone employed in exactly the same way. Mimetic behavior was obviously not mute in the way today's mime is. From the very beginning sounds, gestures and facial expressions would have been vital aids. The standardization and ritualization of the underlying elements in mime would encourage the acting out, within the group, of increasingly involved situations. In such a scenario words would gradually emerge from these mimed accounts, crystallizing into standard sound patterns, a little like the odd letter that will often crop up in a rebus.

In other words: In the beginning there were the stories and, a little at a time, individual words rose out of them. Every single word had to be wrenched free from the narrative context. Language created words and became speech. This is the self-same sequence that is followed with a small child learning to talk. To begin with it has command of a "sentence dictionary," in which words stand for whole sentences; only later are the actual words carved out of these sentence units. As Bronowski asserted: "the normal unit of animal communication, even among primates, is a whole message."[166]

What appeals to me about this model for the origins of language is the way it makes a clear case for two very important aspects of language: First, that language basically puts the stories in our mouths, that it is fundamentally narrative in composition, and fiercely resists being forced to conform to the scientific mind's formal, linear code of conduct.[167] And secondly, that language is every bit as inherently corporeal as it has since proved to be.[168] According to Georg Lakoff the body is the most important aid to understanding the way in which we form concepts. This can be seen, for example, in the way that phrases associated with defeat relate to the body's contact with the ground: "I'm feeling really down," "I'm depressed" or "It's beneath my dignity."

Long before Lakoff, the French philosopher Maurice Merleau-Ponty had highlighted this same observation, albeit within a quite different frame of reference.[169] As far as Merleau-Ponty was concerned, speech and gestures could never express an idea if the body itself was not that idea, rather than its external manifestation. Without an inherent meaning the body would be totally precluded from communicating ideas. Our physical experiences and intentions are inextricably bound up with language.

But to return to Gregory Bateson's theory of the snap as a non-bite. As I said, Bateson suggested that the concept of "not" could have arisen from an "introversion or imitation" of an action. Thus the threat "don't do that," for example, could conceivably have come under the control of the mind in the form of the unadulterated concept of "not."[170]

The relevance of this observation to the point at issue in this chapter is, as I should perhaps remind my readers, that "not" constitutes a puncturing of the space-time continuum which we innocently inhabit and take for granted, inasmuch as it presupposes an alienation, a non-participation—the essence of which is that one is neither that which is denied nor the denial. As such, the "not" concept is—once it is no longer bound to a specifically negative action—no less than the passport to the digital code, to language. And this key would appear to be contained—still unused—within the internal dynamic of the mimetic culture.

Because what the aforementioned aspects of the mimetic culture actually amount to, in essence, is *identification* and, hence, alienation, the fateful split within the incident itself and the identification with it, the image of the incident. The mimetic culture and *Homo erectus* walk the tightrope that separates humanity from its indisputable affinity with nature. On the one side ideas are breaking loose. On the other, ideas are going absolutely nowhere. And perhaps that is where we should have stayed, in the midst of *Homo erectus'* mimetic culture, i.e., at the center of Life's down-to-earth drama.

But we did not, and instead we now find ourselves—in Løgstrup's words—"on the brink of the universe."[171]

In chapter 4 I described the initial process of alienation which occurred on the earth together with the emergence of life, or rather, in the shape of life's inner schism—the split between the organism's analogically coded message and the genetic material's digitally coded message. In chapters 5 and 6 we looked at the way in which this schism gave rise to the development of a semiosphere which permeates the biosphere and penetrates all the way into the tissue deep within us.

And now, in this chapter, we have examined the strange fact that a fresh form of alienation, a split between the analogic reality of experiences and the digital reality of language, came into being at the heart of the semiosphere. The spoken word has endowed the semiosphere with its very own self-referential vertical semiotic system. A new code duality has emerged and with it the dynamic basis for a totally different kind of evolution: cultural history.

When we became human beings, language ran its hyphae far into the nervous system allowing, today, no hope of excision—not even in theory. Language does not think through us but it has become a part of us. And yet language is common property and, hence, extraneous to us.

And this fact—that the spoken word is common property, that it is a tool with which *to share a world* is perhaps the real reason for its emergence. The idea that we all inhabit our very own *umwelt*, an *umwelt* which we take with us to the grave, must gradually have begun to show up on the mental screens of our well-developed, cognitive *erectus* forefathers. At some point it must have dawned on them that they were solitary beings, dissociated from the universe that had engendered them but from which they had broken free by dint of their increasingly emancipated models of the ups and downs of life. The dividing line between things, that fundamental "not," must have begun to have an effect: the recognition of the fact that the line between categories is drawn by "someone" (who can differentiate between A and non-A) and that they too were "someone" and, thus, alien. Because, to become one with the world, "someone" would necessarily have to cease to be "someone."

And this, I believe, brings us to the underlying motivation behind the development of speech. Having been born as self-aware subjects, our forefathers were inevitably also implanted with the seeds of a longing for some greater meaning. The need arose for some kind of mystical fellowship to compensate for the affinity that had been lost. And this is where the collectivization, through speech, of the individual *umwelt*s came to the rescue. Through speech, human beings broke out of their own subjectivity because it enabled them to share one large, common *umwelt*. While pre-lingual creatures had recourse only to their own finite *umwelt*s, speech had the benefit that it could turn the world[172] into a mystically produced common dwelling place. And in creating the myth of the world our forefathers got to grips with a vengeance with the world around them. Language was off and running.

Consciousness: The bodily
governor within the brain

The term *swarm intelligence* captures perfectly the unique aptitude found among social insects such as ants and bees.[173] In an article in the Danish newspaper *Politiken* Benny Lautrup and Claus Emmeche introduced this term in the following fashion: "Ask a neurobiologist about the intelligence quotient of an ant and you will be told that the individual ant is not playing with a full deck. So it seems odd that ants are capable of building colonies as beautifully structured as those we find in the pine forests. What we have here is a collective effect: the colony *en masse* is in possession of a greater intelligence than its individual inhabitants, a swarm intelligence. The individual ant can only carry out very simple tasks, following a scent for instance, and no one ant occupies the post of anthill architect, not even the queen. It is the community as a whole which is the architect. Ants are quite literally the cards in the full deck held by the anthill's higher intelligence. One really ought to say: "Go to the ant*hill,* thou sluggard; consider *its* ways and be wise."[174]

In chapter 7, drawing inspiration from another social insect, the bee, I suggested that animal and plant organisms should be regarded in the same way, that is to say as a *self-organizing chaos* of elements, cells, or pieces of tissue all working their way, more or less independently, to a plan of action that will work for the survival of the organism. And if we turn our attention to an animal's capacity for intelligent behavior, we will discover the aptness here, too, of the term *swarm intelligence*.[175] The chain of command concept to which Western science seems instinctively to defer, because it simplifies the mathematics when the number of dyadic causal connections grows too vast, has had to give way. The brain is, as we have seen, immersed in the immune system's floating morass of physicality and the cognitive scientists' search for the brain's supreme center—or "central processor"—has proved futile. There do not appear to be any such centers or processors. Rather than the brain being pre-programmed to produce intelligence, intelligence seems to swarm out of it.

In order to explain this swarming intelligence "from the top down" models need, for obvious reasons, to be replaced or at least supplemented by "from the bottom up" models, which can be supplied by what are known as "parallel processing computers." These last have unfortunately been dubbed "neural networks," and Søren Brunak and Benny Lautrup describe them as "intuitive computers."[176]

"Parallel" computers do not work on the principle of a logical procedure but, quite simply, through training. Guided by a trainer, the computer gradually improves its performance. By dint of this training it can, for example, become better and better at dividing words correctly. A network has even been produced that can, without the aid of a "trainer"—in other words, by way of a self-organized process—recognize patterns in the input fed into the network. Parallel processing computers are dealt with in more depth in the excellent introduction to Brunak's and Lautrup's book *Neural Networks*.[177]

With the autonomous network a big step may have been taken toward something that might be said to resemble the processes taking place in the real neural network. The implantation of electrodes in those areas of the brain which process olfactory impressions has, for example, revealed the self-organized formation of a specific spatial pattern of activity in rabbits repeatedly exposed to a particular pheromone.[178] The production of this pattern is not, as has traditionally been thought, triggered by the pheromone as such, since the rabbit does not form the pattern until the experiment is repeated. "It is the brain itself that creates the conditions for perception," concludes Christine Skarda, one of the scientists behind this experiment.[179] Perception is a dialogue instigated by the organism on the basis of its own situation; it is not an automatic response brought about by the effect of the receptors on the object.

The relevant aspects of the space-time continuum are not a foregone conclusion. The greatest skill which all living cognition is at all times developing is that of making holes in this continuum, categorizing it, rendering the whole perspective accessible by means of just those fragments of it that are of relevance.[180]

The leap from trainer-based computers to autonomous "neural networks" is, therefore, quite radical, paving the way for a far more inspiring style of model-building. As Claus Emmeche put it: "New pathways must be found that can connect the microscope and the computer. In such a project a degree of 'disciplinary promiscuity' is inevitable. We need periods where one discipline attacks the other; we need exchange and even theft of concepts, methods and perspectives. And to continue our sexual metaphor, we need a dose of interdisciplinary unfaithfulness as well, perhaps some professional mate swapping, some kinky fantasies and some healthy self-criticism that does not degenerate into sadomasochism. At the same time, we ought to be skeptical of any nonreflective combination of various interdisciplinary traditions."[181] The model builders must never forget that the networks are only modelling isolated secondary functions of the body-brain.

Because the true brain is also a body. Embodiment implies a life history, active, uninterrupted participation in life. Human beings and animals are not tucked away behind some screen through which they gain an impression of the external reality on the other side. Quite the contrary. All living creatures are born into the midst of a semiosphere in which they could never acquit themselves were it not for their genetic and biographical experience of it. Until computers start dancing like cranes and mating, while leaving some historic trace of themselves in the future, I have no great faith in them as models for what is going on inside the human head. These so-called neural networks are incomplete or half-finished models—like bicycles that can do no more than wobble and weave downhill because they have neither handlebars nor pedals.

The interplay between the physical and the psychological sides must somehow be introduced into these models. Otherwise we are going to be stuck with dualism. The metaphor of the program or the trainer may mask dualism but it does not get rid of it. Both metaphors merely give the psychological side of dualism a new name, behind which the scientist, on closer reflection, turns out to be the agent (qua programmer/trainer) of that very mind which materialism had forgotten itself enough to think it had risen above.[182]

A journey back through history to trace the origins of swarm intelligence will bring us to the earliest life-forms. Even creatures with no true brain—the sea urchin for example, which has only a simple set of nerve fibers—will invariably execute movements which leave the observer with the feeling that there must be something going on in there. The sea urchin can evade the nasty starfish, which would gobble it up, and some are even capable of differentiating between a hungry starfish and a replete one, fleeing only from the first.[183]

Like all other mammals (and birds) human beings are descended from reptiles. Regarding the vision of these small-brained creatures, Margulis and Sagan have this to say: "Vision in reptiles may be a sort of retinal reflex, a 'response pattern' in which a specific impulse or stimulus leads each time to the same instinctual behavior pattern. Reptiles and amphibians see; but they do not see. If its eye is surgically rotated to produce inverted images the frog never adjusts to the upside-down visual space; the frog unsuccessfully attacks upward with his tongue to catch a fly on his foot."[184] The brain does not intercede as it does among mammals which are capable, after a relatively short space of time, of reversing their "world view" if fitted with the kind of spectacles that turn everything upside down.

Mammal-like reptiles, the so-called synapsids, emerged early on in the Earth's "Age of Reptiles," which stretched from just under three hundred million years ago until around sixty million years ago. But by the close of the Permian period almost all of the synapsids were already extinct. Many may have been eaten by their agile and bloodthirsty cousins, and forerunners of the dinosaurs, the thecodonts. The few surviving species were insignificant little creatures which had become specialists in living their lives under cover of darkness. It is strange to think that the forefathers of today's birds, the dinosaurs (and their forefathers), were therefore our forefathers' worst enemies. For almost two hundred million years our ancestors had to hide from the ancestors of today's birds. And it was only with the mass extinction of the reptiles, which occurred at the end of the Cretaceous Period between sixty and seventy million years ago, that the mammals were really given their chance to see the light of day.

But in the very challenges of this nocturnal existence lies the source of the efficiency that characterizes the swarming intelligence of mammals. For nocturnal creatures, hearing is the most vital sense. They have to be able to pinpoint things by their sound, a feat which calls not only for a well-developed sense of hearing but also some combinative skill, since hearing is not a spatial sense in the same way as sight. Hearing follows time rather than space. Creating an inner chart of the surrounding world necessitates the execution, in the brain, of some pretty exacting computation, whereby something sequential is translated into something spatial. One thing that must occur is a chronological calibration, i.e., a temporal dimension with intervals of appropriate length must be built into the impenetrable arches of the brain. So one might say that the development of the brain in these early mammals hitched a lift from hearing.[185]

When, at long last, the dinosaurs disappeared, and the mammals emerged into the light of day once more, they did not revert to their reptile vision. Instead they developed a far more sophisticated system involving "color reproduction" based on the brain's analysis of visual data supplied by the eye. The French molecular biologist and Nobel prizewinner François Jacob writes: "Visual and auditive information could thus be integrated through a coordinated spatial and temporal code with which it became possible to allocate sound and light stimuli to unique sources, that is, to individual objects that remain constant in time and space. If the brain of higher animals can handle the tremendous amount of information coming in through the sense organs during wakefulness, it is because the information is organized in aggregates, in bodies that constitute the "objects" of the animal's spatio-temporal world, the very elements of its daily experience. Identification and perception of objects can thus be maintained despite changes in the spatial and temporal perception.[186]

In light of this, François Jacob then has this to say on the question of representation: "The external world, the 'reality' of which we all have intuitive knowledge, thus appears as a creation of the nervous system. It is, in a way, a possible world, a model allowing the organism to handle the bulk of incoming information and make it useful for its everyday life. One is thus led to define some kind of 'biological reality' as the particular representation of the external world that the brain of a given species is able to build. The quality of such biological reality evolves with the nervous system in general and the brain in particular."[187] Which, I would venture to add, means evolving with the body as a whole.

Now I am sure that in this term "biological reality" the reader discerns Uexküll's *umwelt*. But since Jacob, like many another modern biologist, seems unlikely to have read Uexküll, the latter receives no acknowledgment.

It is a characteristic of biological evolution that the development of new features is rarely achieved at the expense of the old. Instead, either some sort of modification is made, or a new element will simply be added to the old system. This would appear to be the way in which the development of the brain occurred. The human brain still bears the marks of its reptile past in the form of a "reptile brain," the reptile complex or "R complex," which controls certain deep-rooted response patterns in our lives to this very day. The fact that "the silent hieroglyphic 'thought' of the ritualistic brains of these lizards—assuming that we can become privy to it—illuminates us as well as them"[188] can be illustrated by the squirrel monkey's provocative habit of displaying its erect penis when courting or attacking, when demonstrating its superiority or subservience, and on seeing its own reflection.[189] This behavior is not controlled by the cerebral cortex, nor even by the "mammal part" of the forebrain. But bilateral damage to sections of the old reptile R complex could "short-circuit" this behavioral pattern. "An R-complex-damaged squirrel monkey does not display its penis to its reflection!"[190] Apart from this remarkable alteration in its behavior the R-complex-damaged squirrel monkey seemed surprisingly normal.

If we are to stick with the idea of swarm intelligence, we would perhaps be better to think in terms of swarms of swarms, or perhaps even swarms of swarms of swarms. If that is the case, it is not really a matter of swarms intermingling. In 1985 the neurobiologist Michael Gazzaniga wrote a book entitled *The Social Brain* which took its cue from the notion that: ". . . our brains are organized in such a way that many mental systems coexist in what may be thought of as a confederation."[191]

Models of the brain can easily lead to the absurd conclusion that there is "someone" inside the brain—a tiny homunculus, who is doing our thinking for us. Of course the problem with such models is that inside that homunculus there would have to be yet another homunculus, doing the thinking for the first, and so on. Gazzaniga's concept of the social brain and the idea of swarm intelligence get around this problem because they imply that there is not just some *one* but *many* at home inside the brain. Gazzaniga writes:

> What appears to be a personal conscious unity is the product of a vast array of separate and relatively independent mental systems that continually process information from both the human internal and external environment. Put in more general terms, the human mind is more of a sociological entity than a psychological entity. That is, the human mind is composed of a vast number of more elementary units, and many of these units are capable of carrying out rather sophisticated mental work. [192]

Gazzaniga envisages a situation in which there are hundreds or perhaps even thousands of such brain modules, all of which "express themselves only through real action, not through verbal communication." [193] The majority of these modules are capable of recalling events, storing up emotional responses related to them, and reacting to stimuli connected with a specific memory. All of this activity takes place automatically both in animals and in humans, unconsciously and without recourse to language. Together, these modules provide the fundamental resources for an intellectual life.

In the picture which I am trying to construct each of these modules might be the equivalent of one independent swarm intelligence. But one question then presents itself. Because, if we are to believe Gazzaniga, and we are in fact all running around with possibly thousands of independent brain modules (or thought swarms) inside us, why is it that we experience consciousness as one unified whole? At each and every moment our consciousness seems after all to be encapsulating all the complexity of life on Earth in one single narrative, one "reality," and one self. If it did not, we would actually grow uneasy and begin to feel that something was not quite right. So where does this sense of oneness stem from?

The obvious answer is that all of these brain modules or thought swarms are working together and interacting within one and the same body, a body which is at all times involved in one actual life, one true story. What I am trying to say is that even though consciousness is a neurological phenomenon its unity is a function of the body's own historical oneness. *Consciousness is the body's governor within the brain.* [194]

What happens is this: during every second of a human life, the body is effecting an *interpretation* of its situation vis-à-vis the biographically rooted narrative which the individual sees him- or herself as being involved in at that moment. This interpretation is what we experience as consciousness.

Now at this point the reader might ask how the body can be capable of interpreting, and here I suppose I am, to some extent, at a loss—although I surely have the right, in return, to ask whether the reader thinks it is the brain that does this. And, if so, which of those countless brain modules does the interpreting? Here there is no point in answering that somewhere or other in the brain we have one supreme center. For one thing this does not appear to be the case and for another it merely gives rise to a fresh question: Who ensures that this arch-center can keep track of everything that is going on? Does this arch-center also contain an arch-center? And so on. This is of course a rather unsubtle hint on my part that the reader has tumbled into the homunculus trap. And the reader has no chance of answering back.

It is, at any rate, a fact that consciousness exists as a (discontinuous) sequence of content awareness. I do not believe, as Benjamin Libet has suggested, that it is a body with some sort of right of veto,[195] although I would agree that it might also play that role. I believe that consciousness acts more as a continuous means of taking stock or, as I said, an interpretation.

If we now set aside the question of the interpreter for a while, we might turn instead to the question of what is being interpreted. To me it seems that the answer to this question must be something akin to the total sensory input in which the organism is immersed at any given movement. Every second, all of our hundreds or thousands of brain modules are busily engaged in processing millions upon millions of sensory data, linguistic and otherwise, received from outside or from within. At the same time, a multitude of signs is being exchanged at a furious pace between the brain modules themselves and with muscles and tissues out in the body itself, as "clusters of signs" are hauled in and out of files, i.e., memory banks, which are themselves connected in semiotic loops to all the muscles and glands in the body. This is why pleasant or unpleasant messages are felt within our bodies even before we are consciously aware of receiving them. It actually takes the body almost half a second to generate consciousness.[196] Thus, contrary to what one might automatically assume, consciousness is not one continuous stream but a sequence of discontinuous snippets of content.[197] There might even be spells when it is switched off entirely, though we would never know since one cannot, after all, be aware of one's lack of awareness, far less remember it later.

The whole of this complex, perceptual transformation process, this astronomical swarm of sign processes, constitutes what is known in the Anglo-Saxon world as "the mind"; in this instance let us call it humanity's immediate *umwelt*, our subjective world, that is to say the on-the-spot, conscious or unconscious, construction we put on the significative connections in our lives.

What I suggest is, that those self-same semiotic loops that run between the brain modules and the body, taking care second by second of our bodily functions, are also responsible for the selection process whereby fragments of our *umwelt* are rendered conscious while being integrated into one consciously constructed perception of our participation in a narrative (like this, for instance: here I am, all alone, writing that I am here, all alone . . .). In saying that the body interprets our *umwelt* while generating a constant stream of consciousness, I am thinking of course of the body as one swarming entity, the semiotic brain-body system as a whole. The very fact that we think with our bodies means that consciousness (and language) must be narrative. Physical activity, the elementary act, lies at the root of our intelligence and our consciousness.

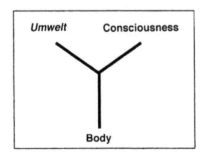

FIGURE 13. *Consciousness as the body's interpretation of its* umwelt.

I would suggest therefore that we look upon consciousness as a purely semiotic relation: *Consciousness is the body's spatial and narrative interpretation of its existential* umwelt.

But, if consciousness is to be viewed thus, as an incessant, self-referential reconstruction of significative connections within an imaginary, narratively arranged mental space, how does this consciousness manage to influence our thoughts and actions? The answer is simple: Consciousness works, as it were, as a switch for switching on or off.[198] On the subject of the "switch" phenomenon, Gregory Bateson wrote:

> The concept "switch" is of quite a different order from the concepts "stone," "table," or the like. Closer examination shows that the switch, considered as part of an electric circuit, does not exist when it is in the on position. From the point of view of the circuit it is no different from the conducting wire which leads to it and the wire which leads away from it. It is merely "more" conductor. Conversely, but similarly, when the switch is off it does not exist from the point of view of the circuit. It is nothing, a gap between two conductors which themselves exist only as conductors when the switch is on. In other words, the switch is not except at the moments of its change of setting, and the concept "switch" has thus a special relation to time. It is related to the notion "change" rather than to the notion "object."[199]

Consciousness is of course not one switch but many—an enormous keyboard. And as the body plays this keyboard of consciousness it switches organs and trumpets, pictures and words, memories and hopes, muscles and glands on and off. The swarming body-brain assumes new shapes which lead to fresh actions, feelings, thoughts, and fancies—and fresh interpretations.[200]

Should the reader now be feeling that our thoughts, rationality, responsibility, and the whole idea of free will have been swallowed up by this hive of activity, my reply would have to be that these phenomena, too, are products of our swarming body-brain, which is extremely susceptible to the playing of the consciousness and exerts tremendous influence on it.

It seems to me that rationality is a hard-won skill which does not come all that naturally to human beings. It would seem to depend on the work of a great many specialist brain modules. This work is presumably being carried out unremittingly—and unconsciously—at some lower level. When we focus our attention on it we summon up its results in our consciousness—couched, possibly, in linguistic terms which act partly as fresh input to the swarms of brain modules and which can also be recalled as a "thought." The actual thought process is, however, something of which we are totally unaware. We could perhaps continue this process by using the results obtained as fresh input, keeping this up until a conclusion seems to have been reached or until we perhaps feel that we have something worth sharing with others.

In order to reach this conclusion we would have to have felt, within our bodies, the thought or train of thought that has vouched for it by preserving it for our minds. At no point, however, is it possible to consciously separate the thoughts from the body.

So, on the one hand, consciousness is a tremendously powerful factor in our intellectual and our active lives and yet, on the other hand, it is quite elusive—determined as it is by the body-brain's capricious interpretation, which can alter from one split-second to the next. Thousands of brain modules are constantly trying to win the attention of the body-brain, like soccer players shouting for the ball. But only the lucky ones whose output the body-brain deems to be of direct relevance to the current "narrative" will gain admission to the consciousness.

That consciousness is, therefore, at the mercy of the body-brain's slightest whim does not prevent the individual from being morally responsible, because of course the individual is not consciousness. Each person's own life story and, hence, personal integrity are at all times present and active within the body-brain. No one entity—not even the consciousness—has the power to make all decisions. And in my view the ego is not automatically the same as the consciousness. I prefer to regard the ego as a biographical category, perhaps something akin to a recollected consciousness, although I would rather not go into that question right here.

When viewed as a process, as a temporal sequence, consciousness might well appear to be deeply involved in the decision-making process. Like swarming bees the body-brain makes its decisions on the basis of an extensive inner communications network within the system. This decision-making process includes a series of "awakenings," as it were, whereby the relevant communication channels are purposefully switched on and off. Major decisions may involve millions of such parallel *and* sequential feedback loops running between the consciousness and the body-brain. The fact that our intelligence is a swarming process in no way prevents it from being both responsible and personal, in the everyday sense of these words. Terms such as these belong to the macro level. They only become significant when applied to the conscious human individual as a unified whole.

"From a distance time seems wide but when one has to pass through it time becomes a narrow doorway," says Princess Irulan in Frank Herbert's novel *Dune*. And likewise, right now, we can imagine every possible kind of future. The spectrum expands rapidly, but once the future has become the present, we all end up living out just the one future. One quadrillion bacteria, in the form of ten trillion cells, collaborate on the job of being a human. Like an astronomical swarm of swarms all of these cells stream together through one single, solitary brain-body as it makes its way along the path of life toward all those unknown futures that will eventually become just one single life story. Uniting is indeed the guiding principle behind this life that lies between birth and death: consciousness' brief illumination of its body-brain. Only in our consciousness do we appear to ourselves as one, or as "someone."

No doubt intelligent animals also have consciousness, and the difference between an animal *umwelt* and a human *umwelt* is probably not so much the presence of consciousness in itself. But the human consciousness is possibly far more restless and mercurial than that of an animal, inasmuch as it is continually being interrupted and steered in new directions due to the cerebral cortex's flexible restructuring of internal and external conditions. The animal's consciousness, on the other hand, is *engrossed* in the realization of its actual life.

I have endeavored, here, to outline a biosemiotic basis for an understanding of the psycho-physical problem, the question of how our psychic life ties in with our biochemical life. This entails a model which, like all other models in this field, is purely speculative. It presents us with a way of transcending dualism, and for that reason I am happy to believe in it.

The idea is that the chemistry of the body, including the electro-chemical phenomena taking place within the brain, is governed by sign processes. As in the case of the aforementioned shot in Sarajevo in 1914—a sign which is in itself almost empty of content and thus, in that sense, dumb—it can instigate events with infinitely greater content. (Since, as we saw earlier, the physicists have commandeered that excellent term *information,* I am obliged here to use the rather vaguer term *content.*) In this way a simple chemical sign, a molecule, can be seized by a hungry receptor. And if a number of receptors all satisfy their hunger at the same time a threshold value may be exceeded. If these receptors happen to be attached to nerve cells that are primed and ready this might be interpreted as a signal to initiate a thought swarm. And all of this is dependent on *the "intelligence" lying not in the sign but in the interpreting body.*

The very fact that the body consists of this infinite swarm of swarming swarms—in which even the smallest unit, the cell, contains a store of historical information that enables it to carry out well-considered interpretations at its own restricted level—makes it possible for intelligent behavior to be induced without any central controlling agency. All that is required is a system of sign processes, an *inner semiosphere.* The exploration of this inner semiosphere ought to be the main aim of modern biology.

One of the inner semiosphere's most interesting means of communication are the *neuropeptides* mentioned at the end of chapter 6. Let us use them as an example in concluding this discussion of human swarm intelligence. Just to recap: neuropeptides are the tiny signal molecules that can be picked up by cells with suitably equipped receptors; and receptors of this sort abound throughout the whole of the body-brain, forming the basis for the communicative network which integrates the brain with the immune system: *the psychosomatic network.*[201]

If we now imagine that the brain is constantly monitoring the notes on the keyboard of consciousness and interpreting the pattern of on-off commands as messages which have to be delivered to specific glands or sections of tissue in the body or brain, then the neuropeptides can be seen as one of many instruments designed to implement these commands. This might, for instance, mean alterations both to the volume of neuropeptides and to the relationship between different kinds of neuropeptides at particular points in the body.

At a conference held in the little town of Tutzing, near Munich in

Germany, the American biochemist Michael Ruff employed the expression "neuropeptide tone" to describe this bodily preparedness of neuropeptides, and he put forward the idea that, since neuropeptides are known to play a part in determining a person's mood and emotional state, one particular state of mind at the biochemical level could conceivably be associated with a specific neuropeptide tone within the body-brain.

This would give us the sort of semiotic relation between consciousness, the nervous system, and the neuropeptide tone shown in figure 14.

What becomes of a cell once its receptor has bound itself to a neuropeptide is dependent not so much on that particular neuropeptide as on what kind of cell it is and what "cell-sociological" and or "cell-historical" state this cell happens to be in. Any given neuropeptide may make one cell move in a particular direction and cause another to alter its form, divide or increase in size, etc. Again we can see that the sign does not embody its own meaning. The meaning must be interpreted by the recipient cell—which gives us a new semiotic relation, as seen in figure 15.

By combining figures 14 and 15 we can see that we have actually come up with a description of a psychosomatic effect as a biosemiotic phenomenon. And, in more general terms, such models provide a tool for depicting psychosomatic illnesses as "unhealthy communication" between cells, tissues, and organs.[203] In recent years this particular psychoimmunoendocrine network has attracted a great deal of atten-

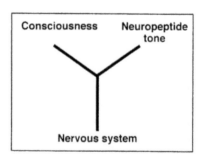

FIGURE 14. *The nervous system interprets the playing of the consciousness as a signal to adjust the neuropeptide tone.*[202]

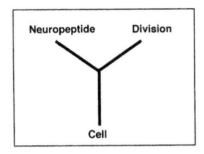

FIGURE 15. *An immune system cell interprets a neuropeptide*
as a signal to commence cell division.

tion, representing as it does one possible path to the understanding of psychosomatic effects.[204]

"If mood is associated with the release of various neuropeptides," Ruff said at the conference in Tutzing, "then the hypothesis of a psychosomatic network implies that all of the cells which participate in the network contribute to the creation of a neuropeptide tone (mood) which then is perceived throughout the body. All illness and disease may result in alterations in neuropeptide tone of the mind-body and such changes may have repercussions for health and well-being."

Still on this subject, Ruff mentioned the fact that in certain types of lung cancer the tumors store and secrete numerous neuropeptides; that these cancer cells are, therefore, components of the psychoimmunoendocrine network and would thus be expected to come under "tonic influence," that is to say, the influence of the neuropeptide tone: "Neuropeptides released in the brain, as well as the body, due to cognitive, emotional or other stimuli may thereby have effects on tumor and/or macrophage growth and tissue localization. These observations may be relevant to the metastatic process *per se,* in that cells which have dissociated from the primary tumor mass invade secondary sites by following gradients of neuropeptide signals."

Ruff concluded his lecture—quite in keeping with a biosemiotic interpretation—by warning against the belief that the growth of a tumor can be checked by treating the patient with drugs which inhibit the effect of one isolated neuropeptide. This kind of "linear" treatment does not take account of the subtle way in which the network operates, or—to use my expression—the complexity of the inner semiosphere.

The skin is no barrier to the semiosphere; it merely marks a shift from one phase to another. The semiosphere meets no resistance as it penetrates to our very core. In a sense our cells are living in the very midst of the semiosphere while we, as conscious individuals, are a kind of epiphenomenon—rather as if the shape of a swarm of bees had the absurd idea that the bees only existed to create that shape.

Granted, throughout evolution brains have had a tendency to make themselves independent of the bodies they served. And this process has of course gone furthest in human beings, where we find not only consciousness but also self-consciousness, i.e., a consciousness that can encompass the sight or thought of itself. This self-referential intelligence has—as we have seen—brought us out of ourselves and introduced us to the community of language. And eventually it began to think that it was something special, that it belonged to a fundamentally different world: while the body had to follow orders the mind was free and full of meaning, or so it thought.

If I seem, in this chapter, to have exaggerated the embodiment of thought, it is because I am taking issue here with this absurd idea, still alive and kicking in our society today, that research and education take their lead from the premises set by it. And no endeavor to overcome this incongruity ever seems to make any headway—logically enough, when one considers that it has to be assessed by experts and boards which draw their own legitimacy from it.

Psychosomatic suffering is the price we pay for this self-propagating folly. Our swarming innards cannot live up to the demand placed on them to be meaningless. The psychological root penetrates deep into the bodily fluids to become illness. The brain-module cell swarms misunderstand one another and send conflicting messages to tissue and glands. We fall ill because our cells cannot quite succeed in uniting to create us.

On ethics: Reuniting two stories
in one body-mind

The schism between nature and humanity is often presented as a distinction between "is" and "ought." Humanity alone occupies an ethical world; only for human beings does an "ought" apply—while nature is in a pure state of being. Yet again it is, of course, Descartes' dualism that is at work here, the idea of things intellectual as being independent and superior to things material. As far back as 1740 the Scottish philosopher David Hume pointed out the logical consequence of this, namely, that one cannot draw valid conclusions from what "is" as to what human beings "ought." Ever since then this prohibition against using scientific discoveries to justify ethical norms has been known as "Hume's guillotine."

Few prohibitions have, however, been violated more often than this one. Scientists and medical men alike have wallowed in the ethical consequences arising from the conquests of science. And they are perhaps not to be blamed for this contempt for Hume's guillotine. For one thing they might not even have been aware of its existence, and for another, ever since Darwin's day it has proved, at heart, to be somewhat ridiculous. Because, if mankind was not created by miraculous means, then it must have been the result of evolution. And thus we are forced to admit that during the course of human prehistory the very thing which is now considered to be a logical impossibility must have occurred: Nature gave rise to man, and "is" became "ought"—all of its own accord.

My critical reader—is she still there?—must by now have figured out that, since I have already dragged language, ideas, and consciousness down into the body's gurgling, rheumy mire, it must now be the turn of humanity's ethical disposition. And that—much as I would have preferred to put it some other way—is exactly what I have in mind. The moment biology is explained in semiotic terms the apparent absurdity is dispelled. Even though the human "ought" appears to be a unique phenomenon in this world, it is nevertheless linked, both logically and biologically, to the alienation of the first living cell, that incompleteness which was established through the duality between the cell and the rewritten version of it in the DNA. From that point onwards, longing's viscous protoform has been a part of our planet. That this proto-longing could, over billions of years of natural history, turn into ethics may well be wonderful—but it is no wonder.

Right at the outset, however, we had better define just what we are talking about here. Through ideas and language every human being inevitably becomes a player in the semiosphere and, in modern society, culture is the predominant feature in humanity's semiotic niche. Since the human race's ability to survive depends to a great extent on humanity's exceptional mental and emotional flexibility, it goes without saying that feelings, thoughts, and values are kneaded and shaped by the cultural conditions under which real people live.[205] So true ethics cannot simply be said to derive from nature.

Few people can have summed up this paradox more wittily than the Danish philosopher Ole Thyssen: "Nature is as generous as a whore. It is on everyone's side and willingly allows itself to be exploited. Nature is not something we find in human beings but something we create in human beings. It is a particular ideal which we use as a tool for interpreting our actions. . . . Villy Sørensen says that the weak see weakness as being inborn and the most human of traits. Anthony Storr says mankind is epitomized by the fact that it has no qualms about kicking someone who is already down. And Freud says that children, who ought to be the closest we can get to innate individuals, are particularly cruel because their capacity for compassion has not yet begun to inhibit their aggressive instincts. Even those who are willing to listen have difficulty in hearing what nature has to say. Academics are divided in their opinions. Their thoughts on nature are in bondage to any and all other thoughts they may have. Thus, the concept of nature is not a suitable basis for any new moral code. Our image of nature is a product of our culture."[206]

But Thyssen is wrong to say that nature is not "something we find in human beings." Although I share his irritation with the arrogant naïveté inherent in determining the nature of mankind from a purely biological point of view, I do not consider it as any kind of progress to deny, out of hand, the idea that humanity's biological roots have something to say about human beings. After all, nature is not alone in being at the mercy of science's cultural prejudice. It actually shares that fate with all other phenomena. And I consider it quite unacceptable to conclude that one might just as well dismantle (or, to use the term currently in vogue, deconstruct) both history and natural history.[207] The fact that our bodies and our culture both play their part in our thoughts does not stop us having them, nor does it prevent their yearning for meaningfulness from having to be satisfied. From the mind's point of view our thoughts may seem laughable but, in that case, surely the same can be said of the very thought that our thoughts are laughable. In its most extreme form "deconstruction" slides into utter fantasy and as such it links up logically with the belief in "artificial intelligence." Both of these bombastic theories forget all about the body and, hence, the unity of natural history.

What we ought to take from all of this is not a sense of hopelessness but above all, and quite simply, humility. Which, in actual fact, means keeping an open mind as regards the perpetual debate over the actual foundations of knowledge. Scholarship is worth no more than the foundation on which it is built, and anyone who does not pay some heed, at regular intervals, to the foundations of their scholarship, is not much of a scholar.

Since, throughout this book, the spotlight has been trained on fundamental issues, I can now launch myself, with perfect equanimity, into this chapter's attempt to establish a semiotically inspired bio-anthropology—that is to say, an account of humanity's biosemiotic origins which may shed some light on our ethical and moral mode of existence.

At what point in evolution did that especial alienation which we call self-consciousness set in? Had our erectus forefathers already learned, one million years ago, to place themselves within their own *umwelt* and see themselves from the outside? In any event this faculty is, as we have seen, closely linked to the discovery of the "not" concept. And it seems likely that a mimetic culture could have been the very source of such an idea. Because, in essence, it is all a matter of being able to replace the actual execution (or mime) of a denial process with an internal impression of it. It must be possible to standardize the threat "do not do that" so that it will eventually manifest itself in the mind's eye as one possible course, but one which need not necessarily be followed. Having got this far, however, yet another mental process is required in order to extract the unadorned "not" from the threat as a whole. It is, in fact, essential that the threatening player is replaceable, i.e., the mental content of the inner image "do not do that" must be so firmly fixed that it will remain virtually unchanged even if the threatening person is replaced.

By this stage self-consciousness must have been ready to make its debut; by this I mean *the idea that the player could be oneself.* And herein lies the root of humanity's unique existential terror. Because this idea brings with it a potential split between the self as an emotional whole and the self as a replaceable, i.e., emotionally neutral, player. A player who is free to contemplate his own mortality.

But we must not forget that this feat, which constitutes the joint birth pangs of both language and self-consciousness, is totally dependent on the ability to empathize with the other. One curious example of this relation can be seen in a small child learning the words "mine" and "yours." The child lying on its changing table says, "that's *my* Teddy." "Yes," says the father, "that's *your* Teddy." But that is of course ridiculous: if the Teddy is mine and Daddy agreed with that, then it can't also be yours. The only way in which the child can clear up this mystery is by switching the players, in other words by seeing the words "your Teddy" from its father's point of view. The child must, therefore, be capable of empathizing with "the other" even before it can talk.

The interchangeability of the players implies an emotional and cognitive relationship between them. Human beings could never have learned to put themselves in someone else's psychological place if they had not already learned to see themselves reflected in that other person, to see the other person as a creature just like them.

The terror and the empathy go hand in hand and language is their medium. In actual fact it was not humanity, much less *erectus,* who engendered them. It was a little child. Lacan's reflection theory holds the key: the mutual empathy between mother and child provided the protection necessary to cope with the unleashing of the awful isolation inherent in the idea of "not."

So the ability to identify and empathize, and—through the common bond of speech—construct a world may be regarded as humanity's extraordinary compensation for the pinprick made by alienation and the recognition of their own mortality in their unthinking symbiosis with the universe. As Arne Næss writes, "An identification process can be defined as a process whereby another being's interests are instinctively responded to as though they were one's own interests."[208] One might even say that we banish the loneliness engendered by this awareness of our mortality by instinctively taking responsibility for one another.[209]

I believe the ethical drama of the human race to be inextricably linked to these existential conditions. Ethics is not about values that we opt for, or that are imposed on us from outside—let alone above; it is about self-knowledge, i.e., the recognition of our ability to empathize as the very lifeline that can help us overcome alienation and the fear of death. The unethical element in persecuting others lies in the fact that we ourselves are brutalized and, thus, betrayed by refusing to listen to our own humane urge to empathize. Because it is through empathy that we become human. And hence, in refusing to empathize, we are rendered inhuman.

But this existential drama also works on yet another level. In becoming articulate beings we lost our innocent grounding, or absorption, in the organic code duality and were swept up instead into the code duality of the linguistic community. Thus our individual life stories became divorced from our genetic history. Or, to put it another way: *Not one but two stories are being enacted in the human body and consciousness.*

This makes it impossible for individuals to fully realize themselves or, in Jung's words, "to individuate" without at the same time seeking to overcome this rift. Each and every one of us has to find some way in our lives of reconciling the two seemingly unrelated stories into which life has plunged us. We must be able to lend an ear to both narratives and unite them into one meaningful story—one which will ensure that we feel at home both in human society and in the living world from which we came.

In actual fact there are three rifts, the first two of which are fundamental and existential. But the third can, at least fleetingly, be healed. We share one of these rifts with all other living creatures, since it relates to the organism's digitally coded self-description in the form of DNA. This rift led to the emergence of living subjects and prompted the narrative which we know as natural history. The second rift is one which we share with all other human beings, but not with animals or plants, and it has to do with the fact of our becoming self-conscious subjects. This rift introduced us into the narrative we know as cultural history.

The third rift is essentially different from the first two, relating as it does to the fact that human beings are involved in both of these great narratives at once without having any idea of how they are connected. Because, in becoming self-conscious subjects, we were sent reeling into the willful labyrinths of culture where the body's slimy, mollusc trail is all too easily lost to view.

With the third rift, too, the remedy must lie in listening to that profound need for empathy. Not, this time, with human beings but with all the life-forms of the earth in general.[210] I wrote earlier that, while straddling the boundary between the mimetic and Stone Age cultures, our forebears must have succeeded in learning how to put themselves in someone else's psycho-logical place. By psycho-logic I mean the narrative logic which controls an event or a story. So our forebears learned to occupy the same position in the narrative, psycho-logical, relational pattern that other human players might conceivably occupy.

By the same token, however, it can be argued that our forebears seem likely to have instituted a bio-logic, a cognitive model of events and stories in which other types of living creatures were accorded interchangeable places; and that, in a given situation, human beings could put themselves in these places. In all probability the gap between human beings on the one hand and other life-forms on the other has been much narrower among these *erectus* forebears in whose body-brains mimetic behavior and the empathic faculty were gradually being kneaded together to form the language of *homo sapiens*. Here, too, the interchangeability of the players must imply that they are emotionally and cognitively related. Our forebears could never have effected this bio-logical process without being able to empathize with animals.

In Gregory Bateson's words, "(What) pattern connects the crab to the lobster and the orchid to the primrose and all four of them to me? And me to you?"[211] This "pattern that connects" is binding in a different way from mere zoology or botany: "Is our reason for admiring a daisy the fact that it shows—in its form, in its growth, in its coloring, in its death—the symptoms of being alive? Our appreciation of it is to that extent an appreciation of its similarity to ourselves."[212]

Bateson's search for the "pattern that connects" went hand in hand with his fierce opposition to dualism. The egoism of Western civilization, its isolation and hate and its economic problems were, to Bateson's mind, the logical consequence of dualism—an opinion which he expressed most pointedly in a lecture given in 1970:

> *If you set God outside and set him vis-à-vis his creation and if you have the idea that you are created in his image, you will logically and naturally see yourself as outside and against the things around you. And as you arrogate all mind to yourself, you will see the world around you as mindless and therefore not entitled to moral or ethical consideration. The environment will seem to be yours to exploit. Your survival unit will be you and your folks or conspecifics against the environment of other social units, other races and the brutes and vegetables. If this is your estimate of your relation to nature and you have an advanced technology your likelihood of survival will be that of a snowball in hell. You will die either of the toxic by-products of your own hate or, simply, of overpopulation and overgrazing. The raw materials of the world are finite.*[213]

But of course dualism does not just involve a split between mind and body, it also entails a hierarchy. The mind gives the orders, the body (i.e., nature, the patient, the pupil) does as it is told.

And the justification of this concept of control may be largely responsible for leading our understanding of the world off course—whether we are dealing with the belief that the Western world can control the rest of the world, or that the government can control society, or that the doctor can control the patient, or that the teacher can control the pupil's learning process, or that human beings can control nature, or that natural selection can control evolution, or that

HEALING

135

the brain can control the body, or that the DNA can control embryogenesis. In every one of these cases dualism is crouched, ready to pounce; because each instance presupposes rationality—i.e., something akin to reason—the logic of which is foisted onto something which in itself is regarded as amorphous and ignorant.

But in every instance we prevent these two sides from entering into an internal and binding exchange with one another. Each time, we consolidate the unbridgeable gap between these two great tales, both of which are working away inside our minds every second of our lives: our cultural history and our genetic history.

If we are to mend this rift, if we are to be *healed,* the only way this can be achieved is through the acceptance of the fact that our intelligence swarms out of our bodies, that we are flesh and blood to the core, and that we occupy a position in our planet's semiosphere which brings us into elementary association with all other life-forms. Empathizing with other life-forms enables us to reconcile, at a symbolic level, our two inner subjects, the anthroposemiotic and the biosemiotic selves.[214] Inside the body, as we saw in the last chapter, psyche and soma are bound to meet. Both our intellectual and our physical health are dependent on these two opponents being transformed into allies in the story of our lives.

Over the past few decades many attempts have been made to safeguard animals from ill-treatment by according them a kind of moral status similar to that of human beings. According to Kant, however, only rational beings, i.e., people, have absolute worth in the sense that they are a "goal in themselves" and, hence, irreplaceable. But this makes no clear allowance for mentally deficient or brain-damaged individuals, nor for infants or fetuses. So they do not benefit from the protection provided by other people's moral obligation to them. Many philosophers have therefore asserted that potentially rational beings, like, for instance, fetuses, also have absolute worth in the Kantian sense. This still does not offer any protection to the animals however; so, in order to bring them into the fold, attempts have been made to establish less stringent criteria.

On this point, Peter Kemp has suggested that any evaluation of which individuals have a right to moral protection must work from the principle of the sanctity of life, and from the value of what could be irrevocably lost. He further maintains that the only way to assess this value is by contemplating what the loss of someone we know would mean to us.[215] We cannot, he says, "approach the natural world as such in the same way as we would approach another person who places their life in our hands."[216]

To Kemp, each person's responsibility to "the Other"[217] is the keynote in any moral stance, and he therefore rejects, in no uncertain terms, the idea that science and, hence, bioanthropology can be used to justify such a stance. It is, however, worth noting that the form of bioanthropology outlined here does lead us to the very sense of responsibility which Kemp demands—albeit not as a humane obligation but as a means of gaining personal peace of mind. And when all is said and done, it is not easy to keep these two motives apart.

I am not entirely sure on what basis Kemp would extend the "idea of the sanctity of human beings" to cover "the idea of the sanctity of all natural life." He rejects—quite rightly, in my opinion—Peter Singer's idea that animals have interests (e.g., in avoiding pain) in the same sense as humans and that they therefore have a right to be taken into account.[218] Such a viewpoint immediately embroils us in a sentimental faith in comparability. Very few things in this world can be compared with one another, without something essential being lost. The fact that I can—with a bit of effort—empathize with a crow, does not mean that a crow's trials and tribulations can, in any way that is relevant, be compared to my own. And, as my wife, Bodil Bredsdorff, says, more or less in jest: "Who can know the scream of a carrot being pulled out of the soil?"

Instead Kemp points out that "the human personality has its roots in the development of an embryo and a body and, hence, in the development of the natural world."[219] And obviously we protect our roots. But is it really so evident that this is done for the roots' sake rather than our own? Yes indeed, replies Kemp, because inasmuch as humanity, with its insight, is superior to the animals, people have the chance to obtain an overall view of nature and, thus, to form a concept of responsibility for the whole of the natural world which they inhabit.[220]

This viewpoint could be neatly summed up as follows: Humanity does not have its roots in responsibility, but human beings are responsible for their roots. That being the case, bioanthropology is obviously an alien element as far as Kemp's ethics are concerned, and this then turns humanity's unique sense of responsibility into an alien element in the natural world.

The Norwegian philosopher Jon Wetlesen has adopted a somewhat different approach in an effort to come up with criteria that would justify the attribution of ethical status to animals, plants, and even ecosystems.[221] As his point of departure Wetlesen takes Spinoza's (1632–1677) idea that all living creatures struggle dauntlessly to keep going, i.e., to maintain their own distinct structural and functional characteristics, and that this struggle is what distinguishes each individual creature. Wetlesen then suggests that we use this Spinozan struggle as the foundation for an "analogic extension" of the concept of ethical status. This kind of innate struggle, Wetlesen says, is not only seen among living creatures but also among entire species, ecosystems and in the biosphere as a whole.[222] With this as his basic premise Wetlesen then goes beyond humanitarian ethics and appoints himself spokesman for environmental ethics.

From this book's biosemiotic standpoint the question of an "analogic extension" of the concept of ethical status can be posed in a slightly different way which leads, nonetheless, to a similar conclusion. The essence of Spinoza's dauntless struggle seems, in fact, to be captured by my concept of organic code duality (cf. chapter 4), the gist of which could be styled as *semiotic survival*—i.e., survival by dint of the body's mortality and the onward progress of the message. Code duality is however somewhat less conservative a concept than Spinoza's struggle. It is only *semi-faithful*, constituting as it does, down through the generations, a source both of constant misinterpretations and of mutations in the shape of distortions, oversights, and reshufflings of the inherited text. And it is precisely because of this "semi-faithfulness" that the interplay between the different levels of the interpretative process (the egg's, the organism's, the gene pool's,[223] the lineage's, the ecosystem's), in which each level forms self-referential strange loops with the others, can lead to the establishment of a generative substratum for the evolutionary process.[224] What is more, only in this evolutionary context can we talk of "the present being incorporated in the future."

And finally—to hark back to the definition of a subject given in chapter 4—only in this context can primitive organisms such as amoebas or mealworms be regarded as true subjects.

But, having opted for code duality as the foundation for an analogic extension of the concept of ethical status, we can hardly then go on to assert that the individual amoeba or the individual daisy constitutes an "ethical subject." Such organisms are ethical subjects only at species level. So, while the individual mealworm has no moral right to protection, the mealworm species does.

With the development of the nervous system things grew more complicated. Animals possessing a nervous system do not just have traces of their genetic past stored away in their genetic material, they also harbor immunological and neural traces—some more well-developed than others—impressions of their surroundings that are based on evolutionary experience. Their *umwelt* has involved them in an exchange with the biosemiosphere which permits primitive learning processes. At what point these learning processes can be considered sophisticated enough for us to speak of true subjectivity at the individual level and not just at the species level is very much a matter of opinion. We will probably have to accept the idea of a gradual transition, but sooner or later, in the development of its neural complexity we ought to recognize an animal as an "ethical subject" in its own right.

But, you might ask, what right have I to introduce code duality as an analogic extension of the concept "ethical status"? Or, to put it another way, why should the possession of the faculty for "subjectivity" as defined in this book be good enough grounds for including a given system in the ethical subject category?

I have had my answer to this question ready for a long time, and this is where bioanthropology comes into the picture. This one animal faculty, "subjectivity," must have been the prime mover in forming our distant forebears' ideas on the bio-logical placing of animals. Could animals and human beings swap places in the imaginary narrative scenario which the mimetic culture learned to cultivate? The answer to this question would determine whether human beings can empathize with animals. My guess would be that our forebears' instinct for the subjectivity of animals has been crucial here. While an individual mosquito has rarely occupied any specific human being's place, it has to be said that more sophisticated animals often have.

At a fair guess, humanity's innate ability to empathize with other types of living creatures could be said to follow a graded scale similar to that which we construct for the subjectivity of the animal concerned. And there are certain types of creature, plants or animals with no individual life history with which we are only able to empathize because they represent some unique vital principle. While one might be able to imagine being a mealworm, it would be hard to imagine being one particular mealworm.

To this, I could add that, if one wished to draw up some sort of graded scale, in such a way that certain types of animal were ascribed a higher ethical status than others, then the concept of "semiotic freedom" which I introduced in chapter 5 would seem to provide the relevant criterion. The character of the animal's *umwelt* is what defines the spectrum of positions that an animal can occupy in the bio-logical sphere, its semiotic niche. The more anthropoid its *umwelt,* the greater our empathy with it. This is exactly the state of affairs I was referring to when (in chapter 5) I quoted Charles Hartshorne's remark that it is "astonishing how much musical intelligibility the utterances of birds have for human beings." In that instance I voiced my doubts that this could be explained simply by saying that birds and people had a common need to send a signal to a listener some distance away. There is another element apart from this; we relish the semiotic artfulness of which birdsong is an expression. It reminds us of something about ourselves. We are not birds, but we can hear that the birdsong means something and thus we can empathize with them.

There can be no doubt either that people have no trouble in empathizing with trees. This obviously creates a problem for me, in my attempt to grade different types of living creatures, because trees have no nervous system. But trees can become subjects for us in a very different way. By virtue of their long life histories we view them as pieces of sculpture that provide us with an instant account of one living creature's hold on life—a century, perhaps, of weathering the vicissitudes of time. The tree's semiosis may well be slow and, literally, rooted to the spot, but in the long run its accumulated message will still be great.

In this case, the term *semiotic freedom* also has the advantage that it cannot be clearly quantified and cannot therefore be misused to create a hierarchy of animals according to any one tyrannical scale.

So, in listening to our existential need to belong to both the living and thinking communities of our planet, we instinctively try to empathize with other living, thinking creatures. This we do for the same reason that we are human beings, that we can talk, that we are aware of ourselves. When, for one reason or another, we ignore our need for empathy, we end up hurting ourselves. But this is not something we are born knowing; at best we learn it by living our lives. Ideally, ethics give this experience an outlet that saves us from hurting ourselves. But rules and generalizations never quite seem to fit the bill and they are especially given to going rapidly out of date. For this reason normal, healthy individuals tend to have a rather strained attitude toward ethics, which are wont to become a big stick in the hands of the powers that be. The ethical debate is essential, not in order to arrive at a new moral code but in order to keep reviving what, at the end of the day, ethics is all about—our existential need to empathize with other *umwelt* builders in this weird and wonderful world. And so, as Niklas Luhmann points out, ethics are used to attack morality just as often as they are used in its defense.[225]

This bioanthropologically based set of ethics is blind to society's and the human community's many options and requirements. As we have seen, it cannot be used for drawing up specific rules. It has to be regarded simply as one among many other forms of input to the ethical process in which we are involved at any given time. Nevertheless, at a time when biotechnology and the environment have become such vital social issues, it introduces a viewpoint on the challenges presented by ethics that has hitherto been lacking.

Usually, biology is only brought into the ethical debate in connection with a particular type of question: When does human life begin? Are human beings essentially promiscuous, aggressive, xenophobic creatures? Is variety a prerequisite for ecological health? And it is of course quite reasonable to pose such questions, but in actual fact the answers to them presuppose an insight that falls outside the bounds of the biosphere. An insight that relates to the semiosphere.

No sooner do we adopt a biosemiotic perspective than we are faced with some very different questions. From a biosemiotic point of view life is not something that ever has a beginning. It is a message that has been with us for millions of years and one which is continually being shown in a new light. The key question is more likely to center on the semiotic freedom of the human fetus. It could for example be said that, in its twelfth week, the fetus has a semiotic freedom on a par with that of a sea urchin. And we might well be able to empathize with such a creature, if we put our minds to it, but we are hardly likely to be left emotionally scarred for not doing so! On the other hand, any proper study of the abortion question ought obviously to take into account the semiotic network that exists between the mother, the father, and the fetus, a network which, in turn, is all bound up with our entire cultural tradition. At this level, bioanthropology does not have a great deal to offer.

The question of humanity's innately unsavory nature also acquires a very different sort of worth. Here it is necessary to emphasize that the genetic influence cannot be interpreted as a hierarchic process. Granted, human beings are born with exceptional semiotic skills, but these are hardly rigid enough to account for an aversion to empathizing with foreigners. And as for the question of variety as a prerequisite for ecological health, this becomes a question of semiospheric health. In the case of agriculture, for instance, the semiotic freedom of the farmer's family and that of those who live off their products must also be taken into account in any survey of the system's health.

I am tossing out these examples, not so much in order to convince my reader or to arouse her dislike, as to show that there is some point in allowing nature into the world of signification, as we do when we consider biology in semiotic terms. This does in fact have far-reaching consequences for our general attitude to all the turmoil of this world.

In 1963, when Rachel Carson wrote her best-selling book *The Silent Spring*—a work which was to prompt the setting up of environmental movements worldwide—she could not have chosen a better title. The loss of birdsong is a most telling symbol for the kernel of the sorrow all human beings feel at the destruction of nature.

But this is just where ecology, which many people believe will be our lodestar in the years ahead, is blind. To the ecologists nature is all about biomass, primary production, currents of energy, nutrient cycles, ecosystems, and biotopes. The fact that we can smell, hear, touch, and handle living creatures is not a matter of any great interest to ecologists. The Swedish science historian Thomas Söderqvist described his own progress from birdwatcher to student of biology in Sweden in the sixties thus: "We were no longer romantic naturalists. We were a bunch of collaborators, in league against those fools who could not see that Everything was connected. Our vocabulary underwent a change. There was no more talk of whooper swans, sedge marshes or boggy pools; now it was all biomass, niches and biotopes. The ecosystem was our God, ecology was our church, and the Odum brothers were the angels He had sent down to Earth."[226]

I would guess that, from a personal point of view, the majority of ecologists are sensitive and receptive to the semiotic aspects of nature. But as a science ecology remains true to the dualist tradition and thus it is primarily preoccupied with the purely physiochemical interplay of animals and plants.

Not only does this approach to research cut us off from the kind of— probably crucial—insights into the logic of nature that only the study of the semiosphere can offer, it also perpetuates the demarcation line between town and country that is mainly to blame for the pathetic attitude to nature of many people today, one that wavers between arrogance and romanticism. The semiosphere does not stop at the city boundary. Admittedly it does change radically insofar as traffic and the communication systems of the industrial areas then begin to dominate the semiosphere. The city is actually a kind of semiotic maelstrom. But as such it is still part of the semiosphere and some people—especially younger people—seem positively to thrive in it. And what right does anyone really have to condemn this fascination with the frenzied semiosis of the city?

In short, it is hard to see how ecology can be our guide and mentor in managing nature when it keeps splitting the world up into two distinct sectors—the natural and the cultural—thereby upholding all of the emotional superstructure, all the illusions, that alienate us from nature.

People can put themselves in another's place and, hence, they can conceive of something being good or bad. There is no sense, however, in berating a plant for being bad, as if it had consciously—or, rather, self-consciously—chosen to do harm.[227] Human beings are capable of imagining another's suffering and there are times when they know that they themselves are guilty of inducing suffering in someone else. Evil, too, presupposes an ability to empathize. But this does not place people outside of nature. On the contrary, evil testifies to the way in which our thoughts and ideas can have material consequences in this life. We cannot simply rise above our physicality and float off like some inconsequential little wave in outer space.

The French philosopher André Glucksmann was probably not the first to point out that, invariably "evil is done in the name of good."[228] People rarely, if ever, think of themselves as being evil, and those people who seem in the eyes of others to be the most evil of all may often believe themselves to be moved by the very greatest of good intentions. Are they not fighting for freedom, justice, God, Allah, science, or whatever the truth, in their eyes, may be? Right now, in the former Yugoslavia, all of these excuses for murder are being bandied about. Good and evil are a team that needs to be handled with extreme care.

This same knowledge may well provide literature with its supreme motivation. In a speech addressed to Salman Rushdie on the third anniversary of Khomeni's *fatwa,* Günter Grass was quite clear on this point: "What others cannot do, we will do—isn't that so, Salman: we will tell, and go on telling, the old, old story in a new way. Our stories do not expose, they bring things out into the open. They feed on the humor of failure, not the triumph of being in the right. The storyteller is not on the victor's side; he lives off loss, and the loser—especially the eternal loser—can put his trust in him."[229]

When I first read this passage I felt strangely comforted. One of the strange truths that life forces us to acknowledge comes with the realization of all the dreadful mistakes we have made, even at crucial points in our lives. And what does true literature feed on if not the painful awareness of how easy it is to make a mistake, how often we see others doing so and how often we ourselves have done so.

The tendency to make mistakes lies at the root of all true development in this world. If no errors had ever arisen in the DNA molecules, life would never have progressed beyond the amoeba stage. And if our distant forebears among the apes had not gradually begun to envisage things, the human race would never have come into being. But of all the possible worlds which human beings can envisage, only one can be right—the others must be wrong. Imagination is the creative exploitation of error.

Basically, the entire intellectual operation consists of discovering that one has made a mistake. But this then leaves us very much the wiser. After all, what can we learn from if not our mistakes? Snapping one's fingers at mistakes is the one sure path to stupidity and inhumanity. Take away the right to offend and freedom of speech ceases to exist, wrote Salman Rushdie. In the belief that anything can be so infallible that it is sacrosanct lie the seeds of terrorism.

And why not admit what reason can never get away from anyway: The world is the most wonderful mistake of all. When once you have witnessed a newborn child's eagerness to make this mistake you might as well bow to fate. Once again the world has triumphed over reason. Fallibility is, in actual fact, what this book is all about—signification and fallibility being the two inseparable sides of the same elementary phenomenon. The one thing there is no mistaking is that which leaves not a trace. Nothingness is the word we give to the sinking sense of meaninglessness that is our greatest dread.

It all started with lumps in nothingness. These lumps coalesced into habits, which we know as laws of nature. The habits paved the way for the emergence of life, since it was now possible, thanks to them, to make predictions—a trait unique to living creatures. But there is no denying that mistakes could also be made. Living creatures could die but, if they had not made too many mistakes, they might be lucky enough to survive as messages in the genetic material. Which meant that they could incorporate traces of the present in the future. In time these traces became intertwined and a network of relations arose to form the basis for increasingly sophisticated types of foresight. In the shape of brains and sensory apparatuses, lumps of one sort lay in wait for lumps of another sort. The semiosphere swelled and swelled and grew ever more demanding.

Finally, in the midst of this semiosphere, human beings—complete with all their self-consciousness—emerged. How it should have come about that this self-consciousness could glorify itself to such an extent that it could eventually imagine that nothing else in this world had any real meaning is a tale that will have to be told elsewhere. What I wanted to demonstrate is simply that this idea and all of its destructive side effects are an illusion. We did not invent meaning. This world has always meant something. It just did not know it.

Down at the bottom of Broadway, near Thirty-Fifth Street, I was suddenly brought up short, heaven knows why, by the thought of Sprogø, a little Danish island that lies midway between the larger islands of Zealand and Fyn. Way back in the days just after the war when cars still had to be reversed off the ferry in Nyborg harbor I can remember Sprogø lying there like an enticing symbol of journeys to far-away places. Now here I was standing on an island at the heart of the unofficial capital of the world, thinking longingly of that other island. Just for a moment this symbol of our journey through life loomed larger than the journey itself. The present shows no mercy to the past. Our bodies and our senses crave their due. We wish to live in the present, yet we carry the traces of the past within us. In some respects these traces stretch back over fifteen billion years and, in their inner form, our cells contain information that is at least three billion years old. The arches and vaults of the brain harbor memories going back hundreds of millions of years. And half a million years ago humanity's existential drama was just starting to take shape.

For my own part, I can remember all the way back to Sprogø, almost half a century ago. And in the meantime, with the building of the bridge across the Great Belt, that little island has become a pier of sorts—a role which, in a way, Manhattan also fulfils. Sprogø and Manhattan, signs and piers. The sign itself is a kind of pier, uniting two coastlines—"someone" with "something." The schism has to be there before humanity can be healed. But the longing is here to stay.

Preface

1. After the publication of this book I learned that the Russian-Estonian semiotician Yuri Lotman had introduced the term "semiosphere" in his book *Universe of the Mind: A Semiotic Theory of Culture* (I. B. Taurus, London, 1990).

1. Signifying

2. Mortensen 1989, p. 149.
3. Gleick 1987.
4. Ibid., p. 23.
5. Nielsen, "Et brag af en nyhed," Information, April 28, 1992.
6. Bateson 1955. Reprinted in Bateson 1972, pp. 177–93.
7. Quoted from Buur Hansen 1993.
8. More on this in chapter 4.
9. Anthony Wilden 1980, pp. 185–86.
10. Ibid., p. 183.
11. Bateson 1970. Reprinted in Bateson 1972, pp. 448–65. Extract here from p. 452.
12. Bateson quotes the following from William Blake: "wise men see outlines and therefore they draw them" (Bateson 1979, p. 108).
13. Wilden 1980, p. 186.

NOTES

2. Forgetting

14. Heinberg 1987, p. 269.
15. Ibid.
16. Ibid.
17. Peirce, 1902. See Buchler 1955, p. 125.
18. Ibid.
19. Paraphrased from Christiansen 1988.
20. As previous note.
21. Peirce 1903. See Buchler 1955, p. 62.
22. Peirce 1897. See Buchler 1955, p. 99.
23. Any deviation from coincidence can be due to natural selection in the classic sense, including sexual selection, etc., but often it would appear to depend upon the communicative behavior of each individual, their participation in the "semiosphere" (see p. 87).
24. Edelman 1987, p. 76.
25. Ibid.

26. It was Thomas Sebeok who pointed this out in 1979: "A full understanding of the dynamics of semiosis may in the last analysis turn out to be no less than the definition of life" (Sebeok 1979, p. 26; see also 1986, p. 211).

3. REPEATING

27. Sørensen 1992, p. 13.

28. Ibid., p. 35.

29. Peirce, 1891. See Buchler 1955, p. 318.

30. Prigogine and Stengers 1984.

31. This debate is presented in Pomian (ed.) 1990.

32. Sørensen 1992, p. 32.

33. Eldredge and Salthe 1984. Salthe 1985. To the best of my knowledge the term *boundary conditions* was first introduced into biology in 1968 in an intriguing article by Michael Polanyi (Polyani 1968).

34. Hoffmeyer 1992a. This term is discussed in greater detail in chapter 5.

35. Sonea 1991. See, e.g., Hoffmeyer 1990.

36. Margulis 1981, Margulis and Sagan 1990. For an excellent exposition of the controversy over the various symbiosis theories, see Sapp 1994.

37. The Canadian microbiologist Sorin Sonea has gone as far as to suggest that the bacteria of this world be regarded as a global organism (Sonea 1991).

38. See Eldredge and Salthe 1984, Salthe 1985.

39. This very useful term, which was originally introduced by the German biologist Jakob von Uexküll, will be examined in greater detail in chapter 5.

40. Wilson et al., p. 852.

41. This forms the solid core of James Lovelock's Gaia theory. See, e.g., Lovelock 1989.

42. A recent treatment of this can be found in Salthe 1993.

43. Lewontin 1991, p. 462.

4. INVENTING

44. While I am an opponent of Descartes' rationalism, I have always had great sympathy with his "I think, therefore I am." But this statement is, at heart, problematic as can for example be seen if one writes it as follows: "I think:'therefore I am.'" Compare Bateson and Bateson 1990, p. 102.

45. Hofstadter 1979.

46. Ibid., p. 18.

47. See, for example, Karl Popper 1972, from p. 44 onward.

48. An excellent summary given in Dyson 1987.

49. Stryer 1988, pp. 113–14. See, for example, "Research News," *Science* vol. 256 (1992), pp. 1396–97. For a recent account, see Deduve 1995.

50. This expression stems from Richard Dawkins 1977 but has since been adopted by many other biologists.

51. J. v. Uexküll 1982, p. 65.

52. Cf. Gregory 1987.

53. The French philosopher Maurice Merleau-Ponty arrived at a related insight through a phenomenological argument. Merleau-Ponty observed that a deeper intentionality comes into play even before the explicit or conscious intentionality. Even when we are unaware of it our actions are intentional, because "originally, consciousness is not an 'I think that' but an 'I can'"—Merleau-Ponty 1945, p. 160. This whole debate is discussed in more depth in Hoffmeyer 1995b.

54. In fact it was not only Lamarck who believed this but almost all of his contemporaries around the year 1800 (Burkhardt 1977). That it is Lamarck alone who is now associated with this theory is most unfair (Hoffmeyer 1985).

55. Wicken 1985, Depew and Weber 1985, Brooks and Wiley 1986, Swenson 1989, Kampis 1991, Salthe 1993, Depew and Weber 1995.

56. Richard C. Lewontin 1992.

57. Kierkegaard 1944, vol. 13.

58. Monika M. Langer 1989, p. 129.

59. A more detailed examination is presented in Hoffmeyer 1992a.

5. OPENING UP

60. Seyfarth and Cheney 1992.

61. Wittgenstein 1968, p. 223.

62. Quoted from memory from Gunnar Olsson 1980.

63. Quoted from Sebeok 1979, p. 39.

64. Ibid.

65. Ibid, p. 45.

66. Wilson 1975, p. 238.

67. McFarland 1987, p. 248.

68. Ibid., p. 248.

69. Albertsen 1990.

70. Uexküll 1982, here p. 45.

71. Ibid., pp. 29–31.

72. Ibid., p. 52.

73. Ibid., p. 45.

74. Ibid.

75. Ibid., pp. 34–35.

76. Ibid., p. 71.

77. Ibid., p. 54.

78. Sebeok 1979, pp. 187–207, and Th. v. Uexküll 1982, pp. 1–24.

79. Claus Emmeche 1994.

80. Levins and Lewontin 1985.

81. Emmeche 1994. The role of behavior in evolution has been analyzed in Plotkin 1988.

82. Wicken 1985, Brooks and Wiley 1986, Swenson 1989.

83. Hoffmeyer 1995.

84. Prigogine and Stengers 1984.

85. McShea 1991.

86. Hoffmeyer 1992a.

87. Ibid., p. 109.

88. Nørretranders 1991, chapter 4.

89. Quoted from Nørretranders 1991, p. 110.

90. Ibid.

91. Ibid.

92. Bruner 1990, p. xiii.

93. Gould 1981.

94. Bateson 1972, p. 315.

95. He had, however, mastered this by the time this manuscript was finished!

96. I doubt, however, whether Bateson himself was ever aware of this connection.

97. Sebeok 1976.

98. Sebeok 1976; Uexküll 1980; Uexküll et al. 1993.

6. DEFINING

99. Margulis and Sagan.

100. Maturana and Varela.

101. Ibid.

102. Linder and Gillman 1992, pp. 36–43.

103. Stjernfelt 1992.

104. Lieberman 1991, p. 38.

105. Ibid., p. 46.

106. Ibid., p. 47.

107. Referred to as somatic mutations. See, for example, Jablonka 1992.

108. The transition from biochemistry to what we call categorical perception is discussed by the biochemist Mogens Kilstrup in *Molekylær biosemiotik* (*Applied Semiotics*), Keld Gall Jørgensen (ed.) to be published by Gyldendal, Copenhagen.

109. Edelman 1989. Chandebois and Faber 1983.

110. Chandebois and Faber 1983, p. 9.

111. Edelman 1989, p. 44.

112. Chandebois and Faber 1983, p. 56.

113. Changeaux 1985, p. 217.

114. Young 1992.

115. Ibid., p. 35.

116. Jerne 1984.

117. Boehmer and Kisielow 1991.

118. Ibid.

119. A. Coutinho, L. Forni, D. Holmberg, F. Ivars, and N. Vaz 1984. Inger Lundkvist, A. Coutinho, F. Varela and D. Holmberg 1989.

120. Jerne 1976.

121. Jerne 1984. For a critique of the idea of the immunological self, see Tauber 1994.

122. Margulis and Sagan 1990.

123. Buss 1987.

124. Varela 1991.

125. A. M. Moulin, quoted from Ilana Löwy 1991.

126. Pert et al. 1985.

127. Ibid., p. 820s.

128. Ibid., p. 824s.

129. Quoted from Sagan and Margulis 1991, p. 372.

7. CONNECTING

130. Emmeche 1990, p. 37.

131. Dennet 1991.

132. Fink 1983, p. 145.

133. 1984, p. 33 ff. This concept was introduced by Tor Nørretranders in 1981 (Nørretranders 1987).

134. Searle 1992, pp. 54–55.

135. Michelsen 1992, p. 122.

136. Ibid., p. 131.

8. SHARING

137. Although it is not exactly clear what abuse, in this context, implies. In chapter 10, however, some suggestion of an explanation for this assertion is given.

138. Quoted from memory from Olsson 1980.

139. Løgstrup 1984, p. 109.

140. Ibid.

141. Wittgenstein 1968. p. 47.

142. Henriksen 1992, p. 29.

143. Shevoroshkin 1990.

144. See, e.g., Donald 1991.

145. Sebeok 1987, p. 344.

146. Lyons 1988, p. 147.

147. Ibid., p. 156.

148. Sebeok 1987, p. 347.

149. Ibid.

150. Donald 1991.

151. Ibid., p. 120.

152. Ibid., p. 121.

153. Ibid., p. 171.

154. Dunbar 1988, quoted from Donald 1991, p. 137 ff.

155. Ibid.

156. Donald 1991, p. 149.

157. This case is mentioned in Donald 1991, p. 151.

158. Tulving 1983.

159. Changeux 1985.

160. This survey was published by Merzenik 1987, quoted here from Donald 1991, p. 13.

161. Bruner 1986.

162. Sebeok 1987.

163. Or, perhaps more accurately, the right arm; cf. Reynolds 1991.

164. Lovejoy 1981.

165. Donald 1991, p. 168.

166. Bronowski 1967, p. 385.

167. Cf. Bruner 1986.

168. Lakoff 1987.

169. Merleau-Ponty 1945.

170. Bateson 1972, p. 425.

171. Løgstrup 1984, p. 14.

172. Gadamer 1975, p. 402.

9. Uniting

173. Deneuborg et al. 1992. A swarm is defined as a set of (mobile) agents which are liable to communicate directly or indirectly (by acting on their local environment) with each other, and which collectively carry out a distributed problem solving.

174. Lautrup and Emmeche, article published in the Danish newspaper *Politiken*, August 22, 1992.

175. The comparison between the brain and an ant colony has previously been made by Douglas Hofstadter (1979, pp. 315–16). Michael Ende presented a more horrific version of the same idea in *The Neverending Story* with the monster Ygramul, the Many.

176. Brunak and Lautrup 1989. The term *neural network* seems, unfortunately, to have stuck, despite there being no proof that these computers reflect the typical neural processes in the brain in any essential fashion. There are, on the other hand, reasons for not believing so. Cf., e.g., Jahnsen and Laursen 1990, p. 96.

177. Brunak and Lautrup.

178. Freeman 1991.

179. Skarda 1992.

180. Varela 1992, p. 255.

181. Emmeche 1991, p. 176.

182. See, for example, Searle 1992.

183. McFarland 1987, p. 61.

184. Margulis and Sagan 1992, p. 160.

185. The theory concerning the relationship between hearing and vision as the generator of mammalian development was originally advanced by Harry Jerison (1973).

186. Jacob 1985, p. 114.

187. Ibid., p. 112.

188. Margulis and Sagan 1991, p. 125.

189. Ibid., p. 151.

190. Ibid.

191. Gazzaniga 1985, p. 6.

192. Ibid., p. 29.

193. Ibid., p. 85.

194. Hoffmeyer 1992a, p. 116.

195. Libet 1989.

196. Ibid.

197. Varela 1991, p. 93.

198. Hoffmeyer 1992a, p. 113.

199. Bateson 1979.

200. Thyssen 1992. There is an interesting formal resemblance between this model and Niklas Luhmann's theory on symbolically generalized media, which gives societies a kind of meta-stability. Luhmann (1984) is discussed in Thyssen 1991 and 1992.

201. Pert et al. 1985.

202. Paraphrased from Hoffmeyer 1992a.

203. Of all researchers in this field, Thure von Uexküll is the one who has attached most weight to this connection. See, for example, Uexküll and Wesiack 1988; Uexküll 1995.

204. Ader et al. 1990.

10. Healing

205. See Hoffmeyer 1984, chapter 7.

206. Thyssen 1982, pp. 14–15.

207. In an earlier book *Naturen i hovedet* (Nature in the Mind's Eye), I suggested the term *historic naturalism* as a modest way of dealing with the fundamental problem posed here. See Hoffmeyer 1984, from p. 231.

208. Næss 1990.

209. Hoffmeyer 1992b, p. 85.

210. Hoffmeyer 1993.

211. Bateson 1979, p. 8

212. Bateson 1979.

213. Lecture given by Gregory Bateson entitled *Form, Substance and Difference*. Printed in Bateson 1972, pp. 448–77. This extract from p. 462.

214. See Sebeok (1989) for a more detailed discussion of these terms.

215. Kemp 1991, p. 185 and Kemp 1992a.

216. Kemp 1991, p. 185.

217. In this case the expression "the Other" should be understood in the light of French philosopher Emmanuel Lévinas' philosophy. See Kemp 1992.

218. Singer 1990.

219. Kemp 1991, p. 185.

220. Ibid., p. 186.

221. Wetlesen 1993.

222. Ibid.

223. Here I would actually have preferred to use the term *genomorph*, which means "genotype pool" rather than "gene pool." See Hoffmeyer and Emmeche 1991, p. 157.

224. Hoffmeyer 1993.

225. See Thyssen 1992.

226. Söderqvist 1991.

227. In Ejvind Larsen's words. Larsen 1991, p. 186.

228. Interview in Danish newspaper *Weekendavisen*, January 27, 1989.

229. Grass 1992.

Ader, R., D. Felten, and N. Cohen, eds. 1991. *Psychoneuroimmunology.* 2d ed. Academic Press.

Albertsen, Leif Ludwig. 1990. "Hvor Kommer Ordet Omverden fra?" *OMverden,* 39.

Anderson, Myrdene, and Floyd Merrell, eds. 1991. *On Semiotic Modeling.* Berlin: Mouton de Gruyter.

Bateson, Gregory. 1972. *Steps to an Ecology of Mind.* New York: Ballantine.

Bateson, Gregory. 1979. *Mind and Nature: A Necessary Unity.* New York: Bantam.

Bateson, Gregory, and Mary Catherine Bateson. 1987. *Angels Fear: Towards an Epistemology of the Sacred.* New York: Macmillan.

Boehmer, Harald von, and Pawel Kisielow. 1991. "How the Immune System Learns about Self." *Scientific American* (October): 50–59.

Bronowski, J. 1967. "Human and Animal Language," in Thomas A. Sebeok, ed., *To Honor Roman Jakobsen: Essays on the Occasion of His Seventieth Birthday,* 375–394. The Hague: Mouton.

Brooks, Daniel, and E. O. Wiley. 1986. *Evolution as Entropy: Toward a Unified Theory of Biology.* Chicago: University of Chicago Press.

Brunak, Søren, and Benny Lautrup. 1988. *Neurale netværk. Computere med intuition.* Copenhagen: Munksgaard. English title: *Neural Networks: Computer with Intuition.* Teaneck, N.J.: World Scientific Publications.

BIBLIOGRAPHY

Bruner, Jerome. 1986. *Actual Minds, Possible Worlds.* Cambridge: Harvard University Press.

Bruner, Jerome. 1990. *Acts of Meaning.* Cambridge: Harvard University Press.

Buchler, Justus. 1955. *Philosophical Writings of Peirce.* New York: Dover Publications.

Burkhardt, R. D. 1977. *The Spirit of System: Lamarck and Evolutionary Biology.* Cambridge: Harvard University Press.

Buss, Leo. 1987. *The Evolution of Individuality.* Princeton: Princeton University Press.

Chandebois, Rosin, and Jacob Faber. 1983. *Automation in Animal Development.* Basel: Karger.

Changeaux, Jean-Pierre. 1985. *Neuronal Man, the Biology of Mind.* Oxford: Oxford University Press.

Christiansen, Peder Voetmann, ed. 1988. *Mursten og Mørtel til en Metafysik.* Roskilde.

Coutinho, A., L. Forni, D. Holmberg, F. Ivars, and N. Vaz. 1984. "From an

Antigen-Centered, Clonal Perspective of Immune Responses to an Organism-Centered, Network Perspective of Autonomous Activity in a Self-Referential Immune System." *Immunological Reviews* 79: 151–168.

Dawkins, Richard. 1976. *The Selfish Gene*. 2d ed. 1989. Oxford: Oxford University Press.

de Duve, Christian. 1995. *Vital Dust: Life as a Cosmic Imperative*. New York: Basic.

Deneubourg, Jean-Louis, Guy Theraulaz, and Ralph Beckers. 1992. "Swarm-Made Architectures," in Fransisco J. Varela and Paul Bourgine, eds., *Toward a Practice of Autonomous Systems: Proceedings of the First European Conference on Artificial Life, Paris*, 123–133. Cambridge: MIT Press.

Dennet, Daniel C. 1991. *Consciousness Explained*. London: Allan Lane, Penguin.

Depew, David, and Bruce Weber, eds. 1985. *Evolution at a Crossroads: The New Biology and the New Philosophy of Science*. Cambridge: MIT Press.

Depew, David L., and Bruce H. Weber. 1995. *Darwinism Evolving: Systems Dynamics and the Genalogy of Natural Selection*. Cambridge: Bradford/MIT Press.

Donald, Merlin. 1991. *Origin of the Modern Mind: Three Stages in the Evolution of Culture and Cognition*. Cambridge: Harvard University Press.

Dunbar, R. I. M. 1988. *Primate Social Systems*. London: Croom Helm.

Dyson, Freeman. 1985. *Origins of Life*. Cambridge: Cambridge University Press.

Edelman, Gerald. 1989. "Topobiology." *Scientific American* (May): 44–52.

Edelman, Gerald M. 1987. *Neural Darwinism: The Theory of Neuronal Group Selection*. Oxford: Oxford University Press.

Eldredge, Niles, and Stanley N. Salthe. 1984. "Hierarchy and Evolution," in Richard Dawkins and Mark Ridley, eds., *Oxford Surveys in Evolutionary Biology*, 184–208. Oxford: Oxford University Press.

Emmeche, Claus. 1990. *Det Biologiske Informationsbegreb*. Århus: Kimære.

Emmeche, Claus. 1994. *The Garden in the Machine: The Emerging Science of Artificial Life*. Princeton: Princeton University Press.

Fink, Hans. 1983. "Om Naturens Grænser." *Philosophia* 1–2: 130–145.

Freeman, Walther. 1992. "The Physiology of Perception." *Scientific American* (February): 34–41.

Gadamer, Hans-Georg. 1975. *Truth and Method*. New York: Seabury.

Gazzaniga, Michael S. 1985. *The Social Brain*. New York: Basic.

Gleich, James. 1987. *Chaos: Making of a New Science*. London: Cardinal.

Gould, Steven Jay. 1981. *The Mismeasure of Man*. London: Norton.

Grass, Günther. 1992. *Information*, 15.–16./2.

Gregory, R. L., ed. 1987. *The Oxford Companion to the Mind*. Oxford: Oxford University Press.

Hansen, Niels Buur. 1992. *Det Guddommelige Spejl: Om Grundtvigs Opgør med Dualismens Erkendelsesteoretiske Mareridt*. Copenhagen: Rosinante/Munksgård.

Heinberg, Claus. 1987. "Økosystemet, Struktur og Evolution," in Niels Bonde, Jesper Hoffmeyer, and Henrik Stangerup, eds., *Naturens Historiefortællere*, 269. Copenhagen: Gad.

Henriksen, Lars. 1992. "Det Sproghistoriske Traume." *Tidsskrift for Sprogpsykologi* 1: 24–30.

Hoffmeyer, Jesper. 1984. *Naturen i hovedet: Om biologisk videnskab.* Copenhagen: Rosinante.

Hoffmeyer, Jesper. 1985. "Fra Lamarck til Lysenko," in Niels Bonde, Jesper Hoffmeyer, and Stangerup Henrik, eds., *Naturens Historiefortællere*, 173–197. Copenhagen: Gad.

Hoffmeyer, Jesper. 1992a. "Some Semiotic Aspects of the Psycho-Physical Relation: The Endo-Exosemiotic Boundary," in Thomas A. Sebeok and Jean Umiker-Sebeok, eds., *Biosemiotics: The Semiotic Web 1991*, 101–123. Berlin: Mouton de Gruyter.

Hoffmeyer, Jesper. 1992b. "Det Humane Genomprojekt: Etiske Overvejelser," in Det etiske råd, eds., *Gen-vejen. Biologien før og nu*, 84–98. Copenhagen.

Hoffmeyer, Jesper. 1993. "Biosemiotics and Ethics," in Nina Witoszek and Elisabeth Gulbrandsen, eds., *Culture and Environment: Interdisciplinary Approaches*, 152–176. Oslo: Centre for Development and the Environment.

Hoffmeyer, Jesper. 1995a. "The Unfolding Semiosphere," in Gertrudis van de Vijver, Stanley Salthe, and Manuela Delpos, eds., *Proceedings of the International Seminar on Evolutionary Systems.* Vienna: forthcoming.

Hoffmeyer, Jesper. 1995b. "The Semiotic Body-Mind." *Cruzeiro Semiótico*, special issue in honor of Professor Thomas Sebeok.

Hoffmeyer, Jesper, and Claus Emmeche. 1991. "Code-Duality and the Semiotics of Nature," in Myrdene Anderson and Floyd Merrell, eds., *On Semiotic Modeling*, 117–166. New York: Mouton de Gruyter.

Hofstadter, Douglas R. 1979. *Gödel, Escher, Bach: An Eternal Golden Braid.* Middlesex: Penguin.

Jablonka, E., M. Lachmann, and M. J. Lamb. 1992. "Evidence, Mechanisms, and Models for the Inheritance of Acquired Characters," *Journal of Theoretical Biology* 158: 245–268.

Jacob, François. 1982. *The Possible and the Actual.* New York: Pantheon.

Jahnsen, Henrik, and Arne Mosfeldt Laursen. 1990. *Hjernevindinger, Vundet af Ny Forskning.* Copenhagen: Nysyn, Munksgaard.

Jerne, Niels K. 1976. "The Immune System: A Web of V Domaines." *Harvey Lectures* 70: 93–110.

Jerne, Niels K. 1984. "The Generative Grammar of the Immune System." *Science* 229: 1059–1067.

Jerrison. 1973. *Evolution of the Brain and Intelligence.* London: Academic.

Kampis, Goerge. 1991. *Self-Modifying Systems in Biology and Cognitive Science: A New Framework for Dynamics, Information and Complexity.* Oxford: Pergamon.

Kemp, Peter. 1991. *Det Uerstattelige. En Teknologi-Etik.* Copenhagen: Spektrum.

Kemp, Peter. 1992. *Lévinas*. Frederiksberg: Anis.

Kierkegaard, Søren. 1944. *The Sickness unto Death*. London: Oxford University Press. First published in 1849 (in Danish).

Lakoff, Georg. 1987. *Woman, Fire, and Dangerous Things: What Categories Reveal about the Mind*. Chicago: University of Chicago Press.

Langer, Monika M. 1989. *Merleau-Ponty's Phenomenology of Perception: A Guide and Commentary*. Tallahassee: Florida State University Press.

Larsen, Ejvind. 1991. "Hvor Politiker ej Tør Træde," in Eva Andelsselskabet, ed., *Det Rene Svineri: Eva's Aarsrapport 91.*, 176–189. Copenhagen.

Levins, R., and R. C. Lewontin. 1985. *The Dialectical Biologist*. Cambridge: Harvard University Press.

Lewontin, Richard C. 1991. "The Structure of Confirmation of Evolutionary Theory." *Philosophy and Biology* 6/4.

Lewontin, Richard C. 1992. "The Dream of the Human Genome." *New York Review*, May 28, 31–40.

Libet, Benjamin. 1989. "Neural Destiny." *The Sciences* (March-April).

Lieberman, Phillip. 1991. *Uniquely Human: The Evolution of Speech, Thought, and Selfless Behavior*. Cambridge: Harvard University Press.

Lindner, Maurine, and Alfred Gilman. 1992. "G Proteins." *Scientific American* (July).

Lotman, Yuri M. 1990. *Universe of the Mind: A Semiotic Theory of Culture*. London: I. B. Taurus.

Lovejoy, C. Owen. 1981. "The Origin of Man." *Science* 211: 341–350.

Löwy, Ilana. 1991. "The Immunological Construction of the Self," in Alfred I. Tauber, ed., *Organism and the Origins of Self*, 43–75. Dordrecht: Kluwer.

Luhmann, Niklas. 1984. *Soziale Systeme: Grundriss einer allgemeinen Theorie*. Frankfurt am Main: Suhrkamp.

Lundkvist, Inger, A. Coutinho, F. Varela, and Dan Holmberg. 1989. "Evidence for a Functional Idiotype Network among Natural Antibodies in Normal Mice." *Proceedings of the National Academy of Science of US* 86: 5074–5078.

Lyons, John. 1988. "Origins of Language," in A. C. Fabian, ed., *Origins: The Darwin College Lectures*, 141–166. Cambridge: Cambridge University Press.

Løgstrup, K. E. 1984. *Ophav og Omgivelse: Betragtninger over Historie og Natur. Metafysik III*. Copenhagen: Gyldendal. Selections from this work published in English under the title "Source and Surroundings" in *Metaphysics I & II*. Milwaukee: Marquette University Press.

Margulis, Lynn. 1981. *Symbiosis in Cell Evolution: Life and Its Environment on Earth*. San Francisco: Freeman.

Margulis, Lynn, and René Fester, eds. 1991. *Symbiosis as a Source of Evolutionary Innovation: Speciation and Morphogenesis*. Cambridge: MIT Press.

Margulis, Lynn, and Dorion Sagan. 1987. *Microcosmos: Four Billion Years of Evolution from Our Microbial Ancestors*. Boston: Allen & Unwin.

Margulis, Lynn, and Dorion Sagan. 1991. *Mystery Dance: On the Evolution of Human Sexuality*. New York: Summit.

Maturana, Humberto, and Fransisco Varela. 1987. *The Tree of Knowledge.* Boston: Shambala.

McFarland, David. 1987. *The Oxford Companion to Animal Behaviour.* Oxford: Oxford University Press.

McShea, Daniel W. 1991. "Complexity and Evolution: What Everybody Knows." *Biology and Philosophy* 6: 303–321.

Merleau-Ponty, Maurice. 1945. *Phénoménologie de la Perception.* Paris: Gallimard.

Merzenich, M. M., R. J. Nelson, J. H. Kaas, M. P. Stryker, W. M. Jenkins, J. M. Zook, and M. S. Cynaderand A. Schoppana. 1987. "Variability in Hand Surface Representations in Areas 3b and 1 in Adult Owl and Squirrel Monkeys." *Journal of Comparative Neurology* 258: 281–296.

Michelsen, Axel. 1992. *Honningbiens Dansesprog, Signaler og Samfundsliv.* Copenhagen: Nysyn. Munksgaard.

Mortensen, Viggo. 1989. *Teologi og Naturvidenskab: Hinsides Restriktion og Ekspansion.* Copenhagen: Munksgaard.

Naess, Arne. 1990. "Man Apart and Deep Ecology: A Reply to Reed." *Environmental Ethics* 12: 185–192.

Nørretranders, Tor. 1987. *Naturvidenskab og Ikke-Viden.* Aarhus: Kimære.

Nørretranders, Tor. 1991. *Mærk Verden.* Copenhagen: Gyldendal. To be published in 1996 in the UK (Penguin) and in the US (Ballantine).

Olsson, Gunnar. 1980. *Birds in Egg: Eggs in Bird.* London: Pion.

Pert, Candace B., Michael R. Ruff, Richard J. Weber, and Miles Herkenham. 1985. "Neuropeptides and Their Receptors: A Psychosomatic Network." *Journal of Immunology* 135/2: 820s–826s.

Plotkin, Henry C., ed. 1988. *The Role of Behavior in Evolution.* Cambridge: MIT Press.

Polanyi, Michael. 1968. "Life's Irreducible Structure." *Science* 160: 1308–1312.

Pomian, Krzysztof, ed. 1990. *La Querelle du Determinisme.* Paris: Gallimard.

Popper, Karl. 1972. *Objective Knowledge: An Evolutionary Approach.* Oxford: Clarendon.

Prigogine, Ilya, and Isabelle Stengers. 1984. *Order out of Chaos.* London: Heinemann.

Reynolds, Peter C. 1991. "Structural Differences in Intentional Action between Humans and Chimpanzees—and Their Implication for Theories of Handedness and Bipedalism," in Myrdene Anderson and Floyd Merrell, eds., *On Semiotic Modelling,* 19–46. Berlin: Mouton de Gruyter.

Sagan, Dorion, and Lynn Margulis. 1991. "Epilogue: The Uncut Self," in Alfred I. Tauber, ed., *Organism and the Origins of Self,* Dordrecht: Kluwer.

Salthe, Stanley N. 1985. *Evolving Hierarchical Systems: Their Structure and Representation.* New York: Columbia University Press.

Salthe, Stanley N. 1993. *Development and Evolution Complexity and Change in Biology.* Cambridge: MIT Press.

Sapp, Jan. 1994. *Evolution by Association: A History of Symbiosis.* New York: Oxford University Press.

Searle, John R. 1992. *The Rediscovery of Mind.* Cambridge: MIT Press.

Sebeok, Thomas A., ed. 1967. *To Honor Roman Jakobsen: Essays on the Occasion of His Seventieth Birthday.* The Hague: Mouton.

Sebeok, Thomas A. 1976. *Contributions to the Doctrine of Signs.* Bloomington: Indiana University Press.

Sebeok, Thomas A. 1979. *The Sign and Its Masters.* Austin: University of Texas Press.

Sebeok, Thomas A. 1986. *I Think I Am a Verb: More Contributions to the Doctrine of Signs.* New York: Plenum.

Sebeok, Thomas A. 1987. "Toward a Natural History of Language." *Semiotica* 65/3/4: 343–358.

Sebeok, Thomas A. 1989. "Preface," in Benjamin Lee and Greg Urban, eds., *Semiotics, Self, and Society,* vi-xiv. Berlin: Mouton de Gruyter.

Sebeok, Thomas A. 1996. "Semiotics and the Biological Sciences: Initial Conditions," forthcoming.

Sebeok, Thomas A., and Jean Umiker-Sebeok, eds. 1991. *The Semiotic Web 1990.* Berlin: Mouton de Gruyter.

Sebeok, Thomas A., and Jean Umiker-Sebeok, eds. 1992. *Biosemiotics: The Semiotic Web 1991.* Berlin: Mouton de Gruyter.

Seyfarth, Robert, and Dorothy Cheney. 1992. "Inside the Mind of a Monkey." *New Scientist* (January 4): 25–27.

Shevoroshkin, Vitaly. 1990. "The Mother Tongue." *The Sciences* (May-June): 20–27.

Singer, Peter. 1990. *Animal Liberation.* New York: Random House.

Skarda, Christine A. 1992. "Perception, Connectionism, and Cognitive Science," in Francisco J. Varela and Jean-Pierre Dupuy, eds., *Understanding Origins: Contemporary Views on the Origin of Life, Mind and Society,* 265–273. Dordrecht: Kluwer.

Söderqvist, Thomas. 1991. "Fri af Økologerne." *OMverden,* 20–22.

Sonea, Sorin. 1991. "The Global Organism," in Thomas A. Sebeok and Jean Umiker-Sebeok, eds., *The Semiotic Web 1990.* Berlin: Mouton de Gruyter.

Sonea, Sorin. 1991. "The Global Organism," in Thomas A. Sebeok and Jean Umiker-Sebeok, eds., *The Semiotic Web 1990,* Berlin: Mouton de Gruyter.

Stjernfelt, Frederik. 1992. "Categorial Perception as a General Prerequisite to the Formation of Signs?" in Thomas A. Sebeok and Jean Umiker-Sebeok, eds., *Biosemiotics: The Semiotic Web 1991.* Berlin: Mouton de Gruyter.

Stryer, Lubert. 1981. *Biochemistry.* 2d ed. San Francisco: W. H. Freeman.

Swenson, Rod. 1989. "Emergent Attractors and the Law of Maximum Entropy Production." *Systems Research* 6: 187–197.

Sørensen, Villy. 1992. *Den Frie Vilje: Et Problems Historie.* Copenhagen: Hans Reitzels.

Tauber, Alfred I. 1994. *The Immune Self: Theory or Metaphor?* Cambridge: Cambridge University Press.

Thyssen, Ole. 1982. *Den Anden Natur.* Copenhagen: Vindrose.

Thyssen, Ole. 1992. "Ethics as Second Order Morality." *Cybernetics and Human Knowing* 1/1: 31–47.

Tulving, Endel. 1983. *Elements of Episodic Memory.* New York: Oxford University Press.

Uexküll, Jakob von. 1982 [1940]. "The Theory of Meaning." *Semiotica* 42/1: 25–82.

Uexküll, Thure von. 1982. "Introduction: Meaning and Science in Jacob von Uexküll's Concept of Biology." *Semiotica* 42/1: 1–24.

Uexküll, Thure von, ed. 1995. *Psychosomatische Medizin.* 5th ed. Munich: Urban & Schwartzenberg.

Uexküll, Thure von, Werner Geigges, and Jörg M. Hermann. 1993. "Endosemiosis." *Semiotica* 96/1/2: 5–52.

Uexküll, Thure von, and Wolfgang Wesiack. 1988. *Theorie der Humanmedizin: Grundlagen ärtzlicheb Denkens und Handelns.* Baltimore: Urban & Schwartzenberg.

Varela, Francisco J. 1991. "Organism: A Meshwork of Selfless Selves," in Alfred I. Tauber, ed., *Organism and the Origins of Self,* 79–107. Dordrecht: Kluwer.

Varela, Francisco J. 1992. "Whence Perceptual Meaning? A Cartography of Current Ideas," in Francisco J. Varela and Jean-Pierre Dupuy, eds., *Understanding Origins: Contemporary Views on the Origin of Life, Mind and Society,* 235–263. Dordrecht: Kluwer.

Varela, Francisco J., and Paul Bourgine. 1992. "Toward a Practice of Autonomous Systems." In *Proceedings of the First European Conference on Artificial Life.* Paris: MIT Press.

Varela, Francisco J., and Jean-Pierre Dupuy, eds. 1992. *Understanding Origins: Contemporary Views on the Origin of Life, Mind and Society.* Dordrecht: Kluwer.

Wicken, J. S. 1987. *Evolution, Thermodynamics, and Information. Extending the Darwinian program.* Oxford: Oxford University Press.

Wilden, Anthony. 1980. *System and Structure.* New York: Tavistock.

Wilson, Edward, Thomas Briggs Eisner, Winslow R., Richard E. Dickerson, Robert L. Metzenberg, Richard D. O'Brian, Millard Susman, and William E. Boggs. 1973. *Life on Earth.* Stanford: Sinaur.

Wilson, Edward O. 1975. *Sociobiology: The New Synthesis.* Cambridge: Belknap.

Wittgenstein, Ludwig. 1968 [1958]. *Philosophical Investigations.* Oxford: Blackwell.

Young, Stephen. 1992. "Life and Death in the Condemned Cell." *New Scientist* (January 25): 34–37.

INDEX

INDEX

JESPER HOFFMEYER is professor in the Biosemiotics Group at the Institute of Molecular Biology at the University of Copenhagen. He has published six books in Danish on social and philosophical aspects of biology and is a regular contributor to leading Danish newspapers. He has been the editor of two major Danish magazines on science, technology, and society, and is presently a member of the borad for The Centre for Ethics and Law, University of Copenhagen.